NUTRITION AND FEEDING OF FISH

Tom Lovell

Auburn University

An AVI Book
Published by Van Nostrand Reinhold
New York

An AVI Book
(AVI is an imprint of Van Nostrand Reinhold.)
Copyright © 1989 by Van Nostrand Reinhold

Library of Congress Catalog Card Number 88-10894

ISBN 0-442-25927-1

Van Nostrand Reinhold
115 Fifth Avenue
New York, New York 10003

Van Nostrand Reinhold International Company Limited
11 New Fetter Lane
London EC4P 4EE, England

Van Nostrand Reinhold
480 La Trobe Street
Melbourne, Victoria 3000, Australia

Macmillan of Canada
Division of Canada Publishing Corporation
164 Commander Boulevard
Agincourt, Ontario M1S 3C7, Canada

16 15 14 13 12 11 10 9 8 7 6 5 4 3 2 1

Library of Congress Cataloging in Publication Data

Lovell, Tom, 1934—
    Nutrition and feeding of fish/Tom Lovell.
        p.   cm.
    Includes bibliographies and index.
    ISBN 0-442-25927-1:
        1. Fishes—Feeding and feeds.   I. Title.
SH156.L68 1988
639.3—dc19   88-10894

To my graduate students

Contents

Contributors

ARAI, SHIGERU, National Research Institute of Aquaculture, Tamaki, Mie, Japan.

HARDY, RONALD W., Northwest and Alaska Fisheries Center, National Marine Fisheries Service, United States Department of Commerce, Seattle, Washington.

LIM, CHHORN, Agricultural Research Service, United States Department of Agriculture, c/o Oceanic Institute, Makapuu Point, Waimanalo, Hawaii.

PERSYN, AMBER L., Tropical Mariculture Technology, Inc., Crystal River, Florida.

ROBINSON, EDWIN H. Delta Branch Experiment Station, Mississippi Agricultural and Forestry Experiment Station, Stoneville, Mississippi.

Preface

Aquaculture is more than a science in its infancy; it is now recognized as a viable and profitable enterprise worldwide. It will continue to grow and supply an increasingly larger percentage of fishery products consumed because the oceans are inadequately managed and their yield is unpredictable. Supply, price, and quality can be controlled more effectively when fish are cultured under managed conditions, like corn in a field. As aquaculture technology has evolved, there has been a trend toward higher yields and faster growth which has necessitated enhancing or replacing natural foods with prepared diets. In many aquaculture operations today, feed accounts for over half of the variable operating cost. Feeding fish in their aqueous environment takes on dimensions beyond those considered in feeding land animals; the nutrient requirements, feeding practices, and feeding environment are unique for fish. Knowledge on nutrition and practical feeding of fish is essential to successful aquaculture.

This book is intended to be helpful to students, scientists, practicing nutritionists, and aquaculturists. It covers the known nutrient requirements and deficiency effects for various fishes. It discusses nutrient sources and preparation of research and practical feeds. It gives direction for designing and conducting fish nutrition and feeding experiments. Feeding practices for several commercially important fishes representing diverse culture systems (coldwater fish, warmwater fish, crustaceans, pond cultures, and highly artificial cultures) are presented. One book, of course, cannot be all-encompassing in the area of fish nutrition and feeding. Practical feeding of other important cultured species of food and ornamental fishes has not been included nor has a thorough coverage of feeding larval fishes. Fish culture is a dynamic area and new technologies are being introduced continuously; some methods discussed in this book may become obsolete quickly. Nonetheless, the material presented has been thoughtfully selected so that it will be of maximum use to persons whose interests range from general aquaculture to animal nutrition to manufacturing fish feeds.

The author acknowledges the assistance given by the five contributing authors and to Sharon Harper, who typed and organized the manuscript.

The Concept of Feeding Fish

EVOLUTION OF AQUACULTURE

Fish farming is believed to have been practiced in China as early as 2000 B.C., and a classical account of the culture of common carp was written by Fan Lei in 475 B.C. (Villaluz 1953). The Romans built fishponds during the First Century A.D. and during the Middle Ages fishpond building was spread throughout Europe by religious men (Lovell, Shell, and Smitherman 1978). Carp farming in Eastern European countries was popular in the 12th and 13th centuries. In Southeast Asia, fishponds were believed to have evolved naturally along with salt-making in coastal areas; the salt beds were utilized to grow milkfish during the rainy season. This practice was originated by the Malay natives before A.D. 1400 (Schuster 1952). Early interest in fish culture in the United States was carried over from England before 1800 and was concentrated on propagation and culture of trout and salmon.

By early in the 20th century, several forms of fish culture were fairly well established, such as milkfish farming in Southeast Asia, carp polyculture in China (Figure 1.1), carp monoculture in Europe, tilapia culture in tropical Africa, culture of indigenous finfish and crustaceans in estuarine impoundments in Asian and Southeast Asian coastal areas, and hatchery rearing of salmonids in North America and Western Europe. With the exception of salmonid culture, these forms of aquaculture were generally extensive, where the nutrient inputs into the system were restricted or limited to fertilizers and crude sources of foods, and yields were low.

Aquaculture has made its greatest advancements during the latter part of the 20th century. New species are being cultured, new technologies have been introduced, a large research base has been established, and commercial investment is being directed into aquaculture. This development has been referred to as the "blue revolution." Aquaculture is now recognized as a viable and profitable enterprise worldwide. It will supply an increasingly larger percentage of fishery products

Fig. 1.1. Fish culture is practiced in China by methods developed centuries ago. Here, grass is put into the pond to feed grass carp. Other carps in the pond, such as silver and bighead, feed on natural pond foods. Manures, but no commercial feeds, are also added to the pond.
(Courtesy of R. O. Smitherman)

consumed. Supply, price, and quality of marine fish fluctuate considerably because the ocean is inadequately managed and its yield is unpredictable. Supply can be controlled more effectively when fish are cultured under managed conditions, similarly to terrestrial animals. High quality products can be maintained because farmed fish usually reach the processing plant alive.

NECESSITY OF FEEDING FISH

As culture technology has evolved, there has been a trend toward higher yields and faster growth. This has necessitated enhancing the

Fig. 1.2. Without supplemental feeding, annual yields of channel catfish in culture ponds are less than 300 kg per hectare, but with concentrated, pelleted feeds annual yields of 5,000 kg per hectare are obtained.

pond food supply by fertilization, supplementing pond foods with crude or concentrated feed materials, or providing all of the nutrients to the fish in a prepared feed (Figure 1.2). As the fish become less dependent on natural food organisms and more dependent on prepared feeds, the need for nutritionally complete feeds becomes more critical. In highly modified environments, such as net pens, suspended cages, and raceways, nutritionally complete feeds are a necessity.

COMPARISON OF FEEDING FISH AND LAND ANIMALS

Feeding fish in their aqueous environment takes on dimensions beyond those considered in feeding land animals. These include the nutrient contribution of natural aquatic organisms in pond cultures, the effect of feeding on water quality, and the loss of nutrients if feed is not consumed immediately. However, the concept of feeding is the same as that applied in feeding other food animals; to nourish the animal to the desired level or form of productivity as profitably as possible. Thus, application of knowledge on the nutritional requirements of fish and the husbandry of feeding various cultured species is essential to successful aquaculture.

Nutrient Requirements

The nutrients required by fish (finfish and crustaceans) for growth, reproduction, and other normal physiological functions are similar to those of land animals. They need protein, minerals, vitamins and growth factors, and energy sources. These nutrients may come from natural aquatic organisms or from prepared feeds. If fish are held in an artificial confinement where natural foods are absent, such as raceways, their feed must be nutritionally complete; however, where natural food is available and supplemental feeds are fed for additional growth, the feeds may not need to contain all of the essential nutrients.

Notable nutritional differences between fishes and farm animals are the following: (a) energy requirements are lower for fish than for warm-blooded animals, thus giving fish a higher dietary protein to energy ratio; (b) fish require some lipids that warmblooded animals do not, such as omega-3 (n-3) series fatty acids for some species and sterols for crustaceans; (c) the ability of fish to absorb soluble minerals from the water minimizes the dietary need for some minerals; and (d) fish have limited ability to synthesize ascorbic acid and must depend upon dietary sources.

Nutritional requirements of fish do not vary greatly among species. There are exceptions, such as differences in essential fatty acids, requirement for sterols, and ability to assimilate carbohydrates, but these often can be identified with warmwater or coldwater, finfish or crustacean, and marine or freshwater species. The quantitative nutrient requirements that have been derived for several species have served adequately as a basis for estimating the nutrient needs of others. As more information becomes available on nutrient requirements of various species, the recommended nutrient allowances of diets for specific needs of individual species will become more refined.

Feeding Practices

Fish are fed in water. Feed that is not consumed within a reasonable time represents not only an economic loss, but can reduce water quality. Therefore, feed allowance, feeding method, and water stability of the feed are factors that the fish culturist must consider, but that the livestock feeder does not. The culture environment may make valuable nutrient contributions to the fish. For example, most waters contain enough dissolved calcium to provide most of the fish's requirement. For fish that feed low on the food chain, such as some tilapias, the pond environment can be a valuable source of protein, energy, and other nutrients.

FISH VERSUS FARM ANIMALS AS ANIMAL PROTEIN IN HUMAN DIETS

Efficiency

Fish convert practical feeds into body tissue more efficiently than do farm animals. Cultured catfish gain approximately 0.84 g of weight per gram of practical diet, whereas chickens, the most efficient warmblooded food animal, gain about 0.48 g of weight per gram of diet (Table 1.1). The reason for the superior food conversion efficiency of fish is that they are able to assimilate diets with higher percentages of protein, apparently because of their lower dietary energy requirement. Fish, however, do not hold an advantage over chickens in protein conversion; as shown in Table 1.1, poultry convert dietary protein to body protein at nearly the same rate as fish. The primary advantage of fish over land animals is lower energy cost of protein gain rather than superior food conversion efficiency. Protein gain per megacalorie of energy consumed is 47 for channel catfish versus 23 for the broiler chicken.

Unfortunately, a total energy (physiological and fossil) budget for production of protein from freshwater fish culture systems has not been developed as precisely as budgets for terrestrial animal and plant

Table 1.1. COMPARISON OF EFFICIENCY OF UTILIZATION OF FEED AND DIETARY PROTEIN AND ENERGY BY FISH, CHICKEN, AND CATTLE

Animal	Feed composition			Efficiency		
	Protein (%)	Energy (kcal ME/g)	ME-protein ratio (kcal/g)	Weight gain per g of food consumed (g)	Protein gain per g protein consumed (g)	Protein gain per Mkcal ME consumed (g)
Channel catfish	32	2.7[1]	8.5	0.84	0.36	47
Broiler chicken	18	2.8	16	0.48	0.33	23
Beef cattle	11	2.6	24	0.13	0.15	6

[1] Metabolizable energy (ME) estimated from digestible energy.
Source: Adapted from Lovell (1979); National Research Council (1983).

proteins and seafood protein. The fossil energy required to grow channel catfish in ponds has been estimated to be similar to that needed for broiler chicken production (Lovell, Shell, and Smitherman 1978); for example, chickens require heating and ventilation and catfish require pumped water and aeration. Processing methods for channel catfish and broiler chickens are also similar (transport from the production site to a nearby processing site, slaughter, and ice-packing or freezing the dressed carcass). Assuming that the fossil energy requirements for producing and processing catfish and chickens are similar, the lower metabolic energy requirement for protein synthesis by catfish (Table 1.1) makes them a more energy-efficient source of protein. Other land animals require more fossil and dietary energy to produce body protein than chickens.

Nutritional Value

The percentage of edible lean tissue in fish is appreciably greater than that in beef, pork, or poultry (Table 1.2). For example, more than 80% of the dressed carcass of channel catfish is lean tissue; only 13.7% is bone, tendon, and waste fat. The caloric value of dressed fish is less than that of the edible portion of beef or pork. The net protein utilization (NPU) value of fish flesh, 83 (as compared to 100 for egg), is about the same as that of red meat, 80, although the essential amino acid profiles of fish and red meat both reflect high protein quality.

Table 1.2. DRESSING PERCENTAGE AND CARCASS CHARACTERISTICS OF VARIOUS FOOD ANIMALS

Source of flesh	Dressing percentage[1]	Characteristic of dressed carcass			Food energy (kcal per 100 g of edible tissue)
		Refuse[2] (%)	Lean (%)	Fat (%)	
Channel catfish	60	14	81	5	112
Beef	61	15	60	25	147
Pork	72	21	54	26	147
Chicken	72	30	65	9	115

[1] The marketable percentage of the animal after slaughter.
[2] In fish, bones only; in beef and pork, bones, trim fat, and tendons; in poultry, bones only.
Sources: Channel catfish, Lovell (1979); beef, M. A. Browning, D. L. Huffman, and W. R. Jones (1988) and United States Department of Agriculture (1986); pork, T. J. Prince, D. L. Huffman, P. M. Brown, and J. R. Gillespie (1987) and United States Department of Agriculture (1983); poultry, E. T. Moran, Jr., Poultry Science Department, Auburn University.

Fish, as well as other animal flesh, is a fair to good source of all of the nutrients except calcium and vitamins A and C. For example, 8-ounce servings of catfish and hamburger would each provide 100% of the recommended daily allowance (RDA) for an adult male of protein, niacin (vitamin), vitamin B_{12} and phosphorus; 25% to 50% of the iron, zinc, and copper; and about 25% of the vitamins thiamine, B_6, and riboflavin. However, this size serving of fish would contain only 280 calories as compared to 750 calories for a similar portion of hamburger.

LEVELS OF FISH CULTURE

Production of Fish Exclusively From Natural Aquatic Foods

Some fish obtain their food exclusively from plankton. These fish are usually continuous grazers and have mechanisms for filtering and concentrating the suspended animal and plant organisms from the water. An example is the silver carp. Others, such as some of the tilapias, have the ability to feed on plankton, but also feed on bottom materials. The common carp is an efficient bottom feeder. Some fishes, such as grass carp, have herbivorous appetites and consume large quantities of higher aquatic plants. Such fishes have been cultured without artificial feeds, but usually with pond fertilization; mostly in areas outside of the United States. This level of production is most applicable in countries where supplemental feeds are expensive or unavailable (Figure 1.3).

Supplementing Natural Foods with Prepared Feed

This level of fish farming essentially involves taking full advantage of natural aquatic productivity and using prepared feeds as a supplement to increase yield further. Usually with species that will accept supplemental feeds, the additional yield of fish resulting from the additional feeding is profitable. For example, the yield of common carp in fertilized ponds is 390 kg/hectare; the addition of grain or grain byproducts increases yields to 1,530 kg/hectare, and high quality supplemental fish feeds further improve yields to 3,000 kg/hectare (Lovell, Shell, and Smitherman 1978). With channel catfish, yields of 370 kg/hectare are obtained from fertilized ponds. With supplements of high-protein feed, yields of 5,000 kg/hectare are obtained in static ponds.

Fig. 1.3. Livestock and fish are cultured in combination in many countries. Above, the pigs are fed and manure from the pig pen fertilizes the pond to produce food for the tilapia.
(Courtesy of L. L. Lovshin)

Where natural aquatic food may make a relatively small contribution to the total protein and energy requirements of the cultured fish, it can provide essential micronutrients that will allow nutritionally incomplete supplemental feeds to be used. As biomass of fish in the pond increases, however, the fish will become more dependent on the supplemental feed for all nutrients. Channel catfish grown in earthen ponds to maximum standing crops of 2,000 kg/hectare grew normally and showed no deficiency signs when vitamin C was deleted from their feed; however, when fish density was increased to 4,000 kg/hectare, growth was normal but resistance to infection was reduced and subclinical deficiency signs occurred.

Intensive Culture of Fish Under Artificial Conditions

With this system, maximum yield per unit of space and effort is a primary concern and highly concentrated, nutritionally complete feeds are justified. Examples of this type of production are rainbow trout cultured in spring-fed raceways and Atlantic salmon growth in net pens in the sea. Production costs are high and nutritionally complete feeds must be fed.

REFERENCES

BROWNING, M. A., D. L. HUFFMAN, W. R. EGBERT, and W. R. JONES.   1988. Proceedings of Reciprocal Meat Conference, Chicago, IL (In press).

LOVELL, R. T.   1979.   Fish culture in the United States. *Science* 206:1368–1372.

LOVELL, R. T., E. W. SHELL, and R. O. SMITHERMAN.   1978.   *Progress and prospects in fish farming*, 262–290. New York: Academic Press, Inc.

National Research Council.   1983.   *Nutrient requirements of warmwater fishes and shellfishes*. Washington, DC: National Research Council/National Academy of Sciences.

PRINCE, T. J., D. L. HUFFMAN, P. M. BROWN, and J. R. GILLESPIE.   1987.   Effects of ractopamine on growth and carcass composition of finishing pigs. *J. Animal Sci.* Suppl 1, Vol. 65.

SCHUSTER, W. H.   1952.   Proceedings Indo-Pacific Fish Council, Special Publication 1. Southeast Asian Fisheries Development Center, Cloilio, Philippines.

United States Department of Agriculture.   1983.   Composition of foods: pork products. Agriculture Handbook No. 8-10, p. 23.

United States Department of Agriculture.   1986.   Composition of foods: beef products. Agriculture Handbook No. 8-13, p. 41.

VILLALUZ, D. K.   1953.   *Fish farming in the Philippines*. Manila, Philippines: Bookman Co.

2

The Nutrients

ENERGY REQUIREMENTS AND SOURCES

One of the striking differences in nutrition between fish and farm animals is that the amount of energy required for protein synthesis is much less for fish than for warmblooded food animals, as is shown in Table 1.1 in the preceding chapter (see page 6). Fish have a lower dietary energy requirement because they do not have to maintain a constant body temperture; they exert relatively less energy to maintain position and to move in water than do mammals and birds on land; and, they lose less energy in protein catabolism and excretion of nitrogenous wastes than land animals because they excrete most of their nitrogenous wastes as ammonia through the gills.

A dietary excess or deficiency of useful energy can reduce growth rate. Because energy needs for maintenance and voluntary activity must be satisfied before energy is available for growth, dietary protein will be used for energy when the diet is deficient in energy in relation to protein. On the other hand, a diet containing excess energy can restrict food consumption and thus prevent the intake of the necessary amounts of protein and other nutrients for maximum growth. Excessively high energy/nutrient ratios can also lead to deposition of large amounts of body fat. This can be undesirable in food fish if it reduces the dress-out yield and shortens shelf life of the frozen fish; however, it may be desirable in hatchery fish raised for release.

Requirements

Information on energy requirements of fish has been slow to accumulate. In practice and research, fish nutritionists have given priority to meeting the requirements for protein, major minerals, and the vitamins and generally have allowed energy to take care of itself. A deficiency or excess of energy will not have great effect on the health of fish. Also, practical feeds for most species made with commonly available ingredients are not likely to be extremely high or low in energy when the protein requirement is met. For example, a 32% protein catfish feed containing soybean meal (50%), grain (40%), and animal byproduct (8%), plus vitamin and mineral supplements (2%), contains approximately 2.8 kcal of digestible energy per gram; this provides an energy (kcal) to protein (grams) ratio of 8 or 9 to 1, which seems to be near optimum.

Livestock and poultry nutritionists have long recognized the importance of meeting energy requirements in formulating practical feeds. Some feeding tables present protein or amino acid allowances as a function of dietary energy plane; that is, as the energy concentration of the diet increases, the protein percentage increases proportionally. The rationale here is that in ad libitum feeding, energy intake regulates food consumption and thus the amount of nutrients the animal ingests daily. Fish, however, are often not fed ad libitum, so nutrient consumption would be controlled by feed allowance and not energy concentration of the diet.

Feeding experiments were used to estimate energy requirements for channel catfish and common carp. The fish were fed diets containing 24% to 35% protein and several measured or calculated levels of digestible energy. Weight gain was the criterion for optimizing energy requirement. Dietary concentrations of digestible energy per gram of dietary protein that effected highest weight gains for channel catfish ranged from 8.7 kcal to 9.7 kcal (Garling and Wilson 1977; Page and Andrews 1973; Prather and Lovell 1973). In similarly designed experiments using 32% protein diets, the optimum energy/protein ratio for weight gain for common carp was 8.3 kcal/g (Takeuchi, Watanabe, and Ogino 1979). When using feeding trials to determine energy requirements for fish, it is important that the availability of energy in the fed diets be precisely known, that the fish be fed as much as they will consume, that the various diets in the experiment be equal in palatability, and that the composition of gain by the fish be measured.

Mangalik (1986) determined digestible energy requirements for channel catfish of three sizes (1 g, 20 g and 100 g) by feeding the fish as much as they would consume with diets containing various energy and protein concentrations and using protein gain as a measure of growth rate. Daily digestible energy requirement for maximum growth was 16.8 kcal/100 g weight for 1-g to 3-g fish, decreasing to 5 kcal/100 g weight for 100-g to 250-g fish. As shown in Table 2.1, protein requirement changed at almost the same rate as the energy requirement with increase in fish size so that the optimum ratio of digestible energy to protein changed only slightly for fish 3 g to 266 g in size.

Table 2.1. PROTEIN AND DIGESTIBLE ENERGY (DE) REQUIREMENTS BY VARIOUS
SIZES OF CHANNEL CATFISH FOR MAXIMUM PROTEIN SYNTHESIS

Fish size (g)	Protein (g/100 g fish/day)	DE (kcal/100 g fish/day)	DE/Protein ratio (kcal/g)
3	1.64	16.8	10.2
10	1.11	11.4	10.3
56	0.79	9.0	11.4
198	0.52	6.1	11.7
266	0.43	5.0	11.6

Source: Mangalik (1986).

Bioenergetics

Bioenergetics is the study of the balance between energy intake in the form of food and energy utilization by animals for life-sustaining processes such as maintenance, activity, and tissue synthesis. The original source of food energy is the sun: through photosynthesis, chloroplasts in green plants capture radiant energy from the sun and convert it into chemical energy through the synthesis of glucose. This compound serves as the hydrocarbon source from which plants synthesize other organic compounds, primarily carbohydrates, proteins and lipids, which are the primary energy sources for fish and other animals.

The basic unit of heat energy is the calorie (cal), defined as the amount of heat required to raise the temperature of 1 g of water 1° C, measured from 14.5° C to 15.5° C. This unit is too small for most convenient use in nutrition. Thus the kilocalorie (kcal), or 1,000 calories, is more commonly used. The international unit of work and energy, the joule, is also used: 1 joule = 0.239 calories or 1 calorie = 4.184 joules.

Gross Energy.   Energy content of a substance is determined by completely oxidizing the compound to carbon dioxide, water, and other gases and measuring the heat released, which is called the gross energy of the product. This is done with an instrument called an adiabatic bomb calorimeter (Figure 2.1). Gross energy values for several pure compounds and feedstuffs are presented in Table 2.2.

Note that fats (triglycerides) have approximately twice as much gross energy as carbohydrates. This is related to the relative contents of oxygen, hydrogen, and carbon in the compounds. Heat is released

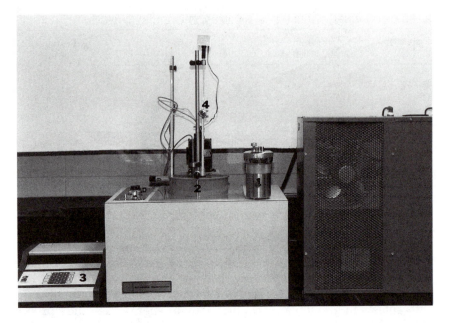

Fig. 2.1. An oxygen bomb calorimeter is used to measure gross energy values of foods. It consists of a bomb (1), in which the food is burned in a concentrated oxygen atmosphere, enclosed in an insulated jacket (2) containing water which absorbs the heat of combustion, controller (3), and mercury (4) or electronic (3) thermometry to indicate temperature rise of water.

only when hydrogen or carbon can react with oxygen from outside of the molecule. Glucose, for example, has enough endogenous oxygen to react with all of the hydrogen in the molecule; therefore, only carbon is oxidized by exogenous oxygen. Because oxidation of a gram of carbon produces only approximately one-fourth as much heat as oxidation of a gram of hydrogen, fats, which have much less endogenous oxygen than carbohydrates, yield more heat upon oxidation than carbohydrates.

Available Energy.   Gross energy content of a food is not a measure of its energy value to the consuming animal. Difference between gross energy and the energy available to the animal for productive purposes varies widely among food materials. Digestibility accounts for most of the differences in available energy among feedstuffs for fishes.

Digestible energy is the difference between the gross energy of the food consumed and the energy lost in the feces. Digestible energy in

Table 2.2. GROSS ENERGY VALUES FOR
SOURCES OF CARBOHYDRATES, FATS,
AND PROTEINS DETERMINED
BY BOMB COLORIMETER

Substrate	kcal/g
Glucose	3.77
Cornstarch	4.21
Triglyceride:	
Beef fat	9.44
Soybean oil	9.28
Casein	5.84

fishes can be determined directly or indirectly. In the direct method, total food consumed and total feces excreted are measured; the indirect method involves collecting only a sample of the food and feces, and digestion coefficients are calculated on the basis of ratios of energy to indicator in the food and feces. An indicator is an inert, indigestible compound in the food; it may be a natural component such as ash or fiber, or it may be an added component such as chromic oxide.

A major consideration in determining digestion coefficients with fish is collecting feces to avoid losses in the water. This means rapid removal of feces from the water before leaching occurs or removal of feces directly from the digestive tract. More detail on methods used in digestion trials with fish are presented by Cho, Slinger, and Bayley (1982) and Lovell and Stickney (1977). Equation (2.1) below is used to calculate digestible energy (DE) by the total collection method and equation (2.2) is used to calculate DE by the indirect method:

$$\% DE = \frac{Food\ energy - Feces\ energy}{Food\ energy} \times 100 \qquad (2.1)$$

$$\% DE = 100 - \left[\frac{Energy\ in\ food}{Energy\ in\ feces} \times \frac{Indicator\ in\ feces}{Indicator\ in\ food} \times 100 \right] \quad (2.2)$$

Metabolizable energy, which represents digestible energy less energy lost from the body through gill and urinary wastes, is more difficult to determine. The fish must be confined in a metabolism chamber to collect gill and urine wastes. The fish are force-fed and total fecal, gill, and urinary wastes are collected. Reliable metabolizable energy values (Table A.1, Appendix A, page 245) have been determined for feedstuffs with rainbow trout; however, channel catfish would not adapt to a metabolism chamber. Metabolizable energy (ME) is calculated according to equation (2.3).

$$\% \ ME = \frac{Food \ energy - (Energy \ lost \ in \ feces, \ urine, \ gills)}{Food \ energy} \times 100 \quad (2.3)$$

Use of metabolizable energy instead of digestible energy to evaluate fish feeds would allow a more absolute estimate of the dietary energy metabolized by the tissues of the animal; also, the National Research Council Committee on Animal Nutrition has adopted this system. Practically, however, metabolizable energy offers little advantage over digestible energy in evaluating useful energy in feeds for fish because energy loss in digestion accounts for most of the variation in recoverable energy among foods. Energy losses through gill and urinary excretions by fish do not vary among foods nearly as much as fecal energy losses and are smaller than nonfecal energy losses by mammals and birds. Furthermore, confinement of the fish in metabolism chambers to determine metabolizable energy is difficult and stresses the fish. Digestible energy is easier to determine and the fish are not stressed when allowed to feed voluntarily. Data in Table 2.3 show that the ratio of digestible energy to gross energy varies greatly among feedstuffs for rainbow trout, but ratio of metabolizable energy to digestible energy varies only slightly. This indicates clearly that digestion accounts for most of the variation in available energy among foods for fish.

Energy Balance in Fish.   Energy flow in fish is illustrated in Figure 2.2. Energy partitioning in fish is similar to that in mammals and birds, but there are several quantitative differences that make fish

Table 2.3. RATIOS OF DIGESTIBLE TO GROSS ENERGY (DE/IE) AND METABOLIZABLE TO DIGESTIBLE ENERGY (ME/DE) FOR RAINBOW TROUT

Feedstuff	DE/IE	ME/DE
Anchovy fish meal	.91	.94
Whitefish meal	.84	.94
Soybean meal, without hulls	.79	.94
Meat meal	.71	.95
Cottonseed meal, without hulls	.63	.93
Wheat middlings	.40	.91

Source: Calculated from values in NRC (1981).

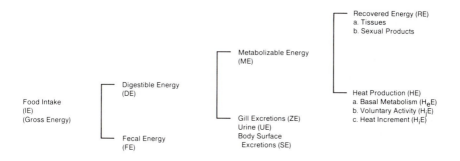

Fig. 2.2.   Partitioning of gross energy in food consumed by fish.

more energy efficient, especially in the assimilation of protein-rich feedstuffs.

Energy losses in urine and gill excretions are lower in fish because approximately 85% of the nitrogenous waste is excreted as ammonia instead of urea (mammals) or uric acid (birds). Heat increment, the rise in energy expenditure associated with the assimilation of ingested food, is lower in fish. Smith, Rumsey, and Scott (1978) measured heat increment by direct calorimetry and found it to be 3% to 5% of ME for rainbow trout; in mammals it may account for as much as 30% of ME. Most of this difference in heat increment is caused by the relatively high energy cost of synthesizing and excreting urea and uric acid as compared with ammonia.

Maintenance energy requirements (voluntary activity and basal metabolism) are lower for fish because they do not have to regulate body temperature and they expend less energy to maintain position in water. Metabolic heat production (kcal/24 h) for small (4 g) rainbow trout was 57 kcal/kg body weight to the 0.63 power (Smith, Rumsey, and Scott 1978) versus 70 kcal to 83 kcal/kg body weight to the 0.75 power for mammals and birds (Brody 1945).

Energy Sources

Because fishes evolved in an aqueous environment where carbohydrates were scarce, their digestive and metabolic systems seem to be better adapted to utilization of protein and lipids for energy than carbohydrates. Some fishes, however, such as warm-water herbivores or omnivores, can digest and metabolize carbohydrates relatively well. Salmonids utilize carbohydrates poorly. Although some fishes consume macrophytes (higher aquatic plants) readily, they digest native cellulose poorly.

Carbohydrates.   The basic chemical structure of carbohydrates consists of sugar units that are aldehyde or ketone derivatives of polyhydric alcohols containing carbon, hydrogen, and oxygen. Hydrogen and oxygen are usually in the same ratio as in water and as in glucose (Figure 2.3). Carbohydrates exist in nature as ringed structures and are more accurately depicted in the Haworth perspective on the right.

Carbohydrates are classified by the number of "sugar" units in the molecule. *Monosaccharides* are one-sugar units, such as glucose (6-carbon) and ribose (5-carbon). *Disaccharides* are conjugates of two monosaccharides. Examples are maltose, composed of two glucose units, and sucrose, composed of glucose plus fructose. *Polysaccharides* are long-chain polymers of monosaccharides. The two most important carbohydrates in animal nutrition are starch and cellulose, each being polymers of glucose units; the difference between the two is the type of glucose molecules. Starch (Figure 2.4) contains α-D-glucose (glucose units are joined by a 1-4 linkage) and cellulose contains β-D-glucose (glucose units are joined by a 1-4 linkage).

Cellulose (Figure 2-5) is the major structural component of plant cell walls. It is highly insoluble at neutral pH and is indigestible to monogastric animals, including fish. Cellulose has a flat, band-like structure, instead of a helical form as starch, and the molecules are held more firmly to each other by hydrogen bonding.

The endosperm of grains is composed mostly of starch and this is the major source in animal feeding. Two forms of starch are found in the starch granules in grains: amylose and amylopectin. Amylose is a straight-chain, α-1,4 glucose polymer. It comprises 20% to 30% of the starch granule and is soluble in warm water. Amylopectin is a branched chain glucose polymer; the α-1,4 straight chain is branched

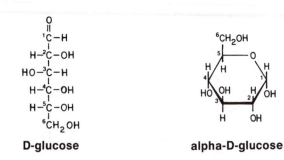

D-glucose   **alpha-D-glucose**

Fig. 2.3.   Glucose.

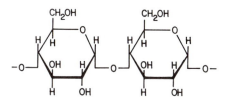

Fig. 2.4.   Starch (α 1-4 linkage).

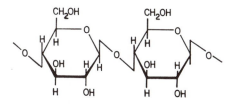

Fig. 2.5.   Cellulose (β 1-4 linkage).

by α-1,6 linkage to a side chain. Starch granules are different in morphology and solubility among plant species. Some are quite resistant to rupture, which is necessary for digestion. Moist heating ruptures, or gelatinizes, the granule and increases solubility and digestibility of the starch. Glycogen is the carbohydrate energy reserve in animal tissue, mainly the liver. It is similar to amylopectin in molecular weight, but its 1,6 linked side chains are shorter and it is more soluble in water.

Although carbohydrates are a significant source of energy and are components of a number of body metabolites, such as blood glucose, nucleotides, and glycoproteins, they are not essential nutrients. Brambila and Hill (1966) showed that chicks can grow normally on carbohydrate-free diets if the calorie/protein ratio is optimum and if triglycerides are included in the diet to supply glycerol for carbohydrate synthesis. Several studies have shown that fish grow satisfactorily and show no pathologies when fed carbohydrate-free diets.

Lipids.   Lipids are a large, varied group of organic compounds that are insoluble in water, but soluble in organic solvents. They represent concentrated energy sources, vitamins, pigments, and essential growth factors for fish. The lipids that are important energy sources are fats, or triglycerides. Chemically, these are esters of fatty acids with glycerol. One mole of glycerol unites with three similar or different fatty acids, with the loss of 3 moles of water (Figure 2.6). R in the

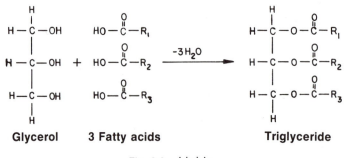

Fig. 2.6.   Lipids.

model represents the hydrocarbon chain in the fatty acid. Length of the carbon chain in most of the fatty acids in land plants and animals is 14 to 18 carbons and in fish, up to 22 carbons.

The chain length and number of double (unsaturated) bonds determines physical and nutritional properties of fats. Fatty acid composition of triglycerides from several sources is presented in Table A.4 in Appendix A (page 253) on Composition of Feed Ingredients. Generally, the fat stores in warmblooded animals are highly saturated (few double bonds), while fats from plants are more unsaturated; however, chain length of fatty acids from land animals or plants is seldom longer than 18 carbons. Fats of farm-raised channel catfish resemble those of grain-fed livestock. Also, salmonids fed diets with saturated animal fats tend to have fats relatively similar to those in the diets. Wild fishes, whether they come from cold or warmwater, freshwater, or marine environments, have significant amounts of polyunsaturated fatty acids 20 carbons and longer in length. They obtain these through the food chain from algae, which synthesize them in the aqueous environment.

Proteins as Energy Sources.   Fish use protein efficiently as a source of energy. A higher percentage of the digested energy in proteins is metabolizable in fish than in land animals. Also, the heat increment for protein consumed is lower in fish than in mammals or birds, which gives protein a higher productive energy value for fish. This apparently is attributed to the efficient manner of nitrogen excretion in fish. As with land animals, excessive amounts of protein in the diet in relation to nonprotein energy suppresses growth rate of fish. Studies with channel catfish showed that increases in dietary protein above 45%, without proportionate increases in nonprotein energy, suppressed growth rate.

PROTEINS AND AMINO ACIDS

Amino acids are the structural components of proteins. The basic structure of an amino acid is illustrated in Figure 2.7. The essential components are a carboxyl group (—COOH) and an amino group (—NH₂) on the alpha carbon atom. All have the basic structure shown where R is the remainder of the molecule attached to the alpha carbon. Amino acids are linked together by a peptide bond to form proteins. Proteins contain carbon (50–55%), hydrogen (6.5–7.5%), nitrogen (15.5–18%; an average value of 16% is assumed), oxygen (21.5–23.5%), and usually sulfur (0.5–2.0%).

There are 18 amino acids that can be found in most any plant or animal protein, although proteins usually contain 22 to 26 amino acids. Amino acids can be conveniently classified into groups according to the series of organic compounds in which they belong. The formulas of 22 are presented in Figures 2.8 through 2.13.

Types of protein found in the fish body are generally based upon function or solubility. *Fibrous proteins* are highly insoluble (indigestible) proteins and include collagen, elastin, and keratin. Collagen is the component of connective tissues, bone matrix, skin, scar tissue, fins, gill operculum, and blood vessels. Elastins are found in arteries, tendons, and other stretch tissues. Keratins are found in hair and hooves of land animals, but in only small amounts in fish. *Contractile protein* is the muscle protein complex. Three proteins, actin, tropomyosin B, and myosin, take part in muscle contraction. Muscle protein is highly digestible and has high nutritional value. *Globular proteins* are proteins extractable from tissue with water or dilute salt solutions. They represent enzymes, protein hormones, and proteins of the serum (soluble) fractions of blood.

Essential Amino Acids

Amino acids can be divided into two nutritional groups, *essential* and *nonessential*. The essential amino acids are those that the animal cannot synthesize or cannot synthesize in sufficient quantity to support maximum growth. The nonessential amino acids are those that can be synthesized by the animal in quantity to support maximum growth. Most monogastric animals, including fish, require the same 10 essential amino acids: arginine, histidine, isoleucine, leucine, lysine, methionine, phenylalanine, threonine, tryptophan, and valine. In the rat, several of the essential amino acids, arginine, histidine, isoleucine, leucine, methionine, phenylalanine, tryptophan, and valine, can

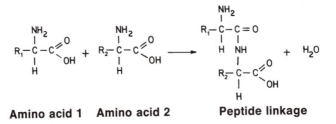

Amino acid 1   Amino acid 2   Peptide linkage

Fig. 2.7.   Basic structure of an amino acid.

  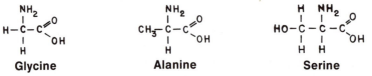

Glycine   **Alanine**   **Serine**

  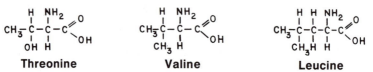

Threonine   **Valine**   **Leucine**

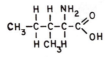

Isoleucine

Fig. 2.8.   Aliphatic series.

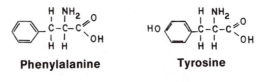

Phenylalanine   **Tyrosine**

Fig. 2.9.   Aromatic series.

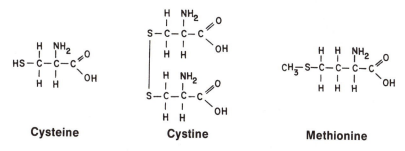

Cysteine **Cystine** **Methionine**

Fig. 2.10. Sulfur-containing series.

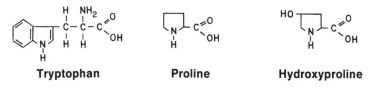

Tryptophan **Proline** **Hydroxyproline**

Fig. 2.11. Heterocyclic series.

Aspartic acid **Glutamic acid**

Fig. 2.12. Acidic series.

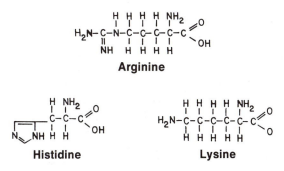

Arginine

Histidine **Lysine**

Fig. 2.13. Basic series.

be replaced by their corresponding α-hydroxy or α-keto analogs, indicating that the carbon skeleton is what the animal is unable to synthesize. However, these analogs for threonine and lysine are not utilized for growth by rats.

Qualitative requirements for amino acids are determined in fish by feeding a purified diet composed of crystalline amino acids as a control diet, and feeding test diets similar to the control except that one amino acid at a time has been removed. Test diets that produce no growth or markedly less than the control represent amino acids that are essential to the fish. Quantitative requirements for essential amino acids are determined by feeding graded levels of one amino acid at a time in a test diet containing crystalline amino acids or a combination of a purified protein and crystalline and amino acids. The amino acid profile of the test diet is usually similar to that in chicken or fish eggs, or of the fish muscle. Growth data from the amino acid feeding trials are equated to amount of amino acid in the diet and the requirement is determined by estimating or calculating the break point in the response curve. In early studies involving salmon, the requirement was estimated by visual inspection. Later, studies with channel catfish used the statistical method of Robbins, Norton, and Baker (1979) to determine the break point in the growth response curve. Santiago (1985) determined two requirements for essential amino acids with Nile tilapia, one for maximum growth (Y_{max}) and one for a level of growth less than maximum (Y_1) but within the 95% confidence limit of Y_{max}. This is possible by fitting the growth response data with nonlinear regressions, as described by Zeitoun, Ulbrey, and Magee (1976). The benefits of the latter method for calculating the amino acid requirements for commercial fish feeds are described in chapter 6, Fish Feeding Experiments.

In some cases, fish can partially substitute a nonessential for an essential amino acid. Channel catfish grow satisfactorily when methionine is the only sulfur-containing amino acid in the diet, but not when cystine is the only sulfur-containing amino acid; however, cystine can replace about 60% of the methionine. Tyrosine, a nonessential aromatic amino acid, can replace about one half of the channel catfish's requirement for phenylalanine, an essential aromatic amino acid.

Dietary imbalances of amino acids can cause reduced performance by animals through amino antagonism or toxicity. When some amino acids are fed in excess of their required levels, they cause an increase in the requirement for other structurally similar amino acids, or *amino acid antagonism*. In some instances, however, dietary excesses of certain amino acids are directly toxic and their negative effects cannot be ameliorated by additions of other amino acids; this is *amino acid toxicity*. Fish diets containing practical feedstuffs, such as grain byproducts, oil meals, and animal byproducts, are not likely to be so seriously imbalanced, but special diets could be.

The quantitative amino acid requirements of five fish species are presented in Table 2.4, along with comparable values for swine and chicken. Except for arginine, the amino acid requirements of fish follow a relatively similar trend as those for other animals. With the exception of arginine and methionine plus cystine, the amino acid requirements of channel catfish are similar to those of chinook salmon. However, the requirements for several amino acids are higher for common carp and Japanese eel. There is probably less variation among fishes than these initial data indicate. It should be pointed out that the requirements presented in Table 2.4 represent usually one study for each fish. Factors such as fish size, temperature, genetics, feeding rate, energy concentration and other diet factors, and method of data analysis can influence the reported requirement for amino acids. In view of the economic importance of amino acid requirements in formulating commercial feeds, these requirements should be re-evaluated.

Meeting Amino Acid Requirements in Practical Feeds

Amino acid requirements of fish are presented in Table 2.4 and in the National Research Council feeding tables on the basis of being 100% available, whereas amino acid composition of feedstuffs is usually presented on a total (not available) content basis. Thus, in formulating fish feeds to meet amino acid requirements, the total amino acid content of the feed ingredients must be corrected for availability (digestibility) to allow the optimum amounts of amino acids in the diet. Digestion coefficients for individual amino acids in several feedstuffs were determined for channel catfish by Wilson, Robinson, and Poe (1981). Digestible amino acid content of these feedstuffs is presented in Table A.2 in Appendix A (page 247). Digestibility of some amino acids varies among feed ingredients; for example, apparent digestibility of lysine is 27% lower in cottonseed meal than soybean meal. Generally, though, the digestibility of the protein may be assumed in estimating the availability of amino acids in the feedstuff when digestibility of individual amino acids is not known.

The research literature is unclear on the efficacy of supplementing fish feeds with isolated amino acids (Lovell 1984). Individual sup-

Table 2.4. ESSENTIAL AMINO ACIDS REQUIREMENTS OF SEVERAL FISHES, CHICKENS, AND SWINE (PERCENTAGE OF THE PROTEIN)

Amino acid	Japanese eel[1]	Common carp[2]	Channel catfish[2]	Chinook salmon[2]	Tilapia nilotica[3]	Chicken[2]	Swine[2]
Arginine	4.2	4.2	4.3	6.0	4.2	5.6	1.2
Histidine	2.1	2.1	1.5	1.8	1.7	1.4	1.2
Isoleucine	4.1	2.3	2.6	2.2	3.1	3.3	3.4
Leucine	5.4	3.4	3.5	3.9	3.4	5.6	3.7
Lysine	5.3	5.7	5.1	5.0	5.1	4.7	4.4
Methionine	3.2	—	—	—	—	—	—
(+ cystine)	5.0	3.1	2.3	4.0	3.2	3.3	2.3
Phenylalanine	5.6	—	—	—	—	—	—
(+ tyrosine)	8.4	6.5	5.0	5.1	5.7	5.6	4.4
Threonine	4.1	3.9	2.0	2.2	3.6	3.1	2.8
Tryptophan	1.0	0.8	0.5	0.5	1.0	0.9	0.8
Valine	4.1	3.6	3.0	3.2	2.8	3.4	3.2

[1]From: Arai (1986).
[2]From: NRC (1979, 1981, 1983, 1984).
[3]From: Santiago (1985).

plementation of soybean meal with lysine, methionine, histidine, or leucine did not improve growth rate of rainbow trout, but collective supplementation did increase growth rate. Methionine supplementation of commercial soybean meal improved growth rate of rainbow trout, but methionine supplementation of reheated soybean meal did not. There was no benefit in supplementing soybean meal in catfish diets with methionine or lysine; however, supplementing soybean meal with both methionine and lysine improved carp diets. Feeding supplemental lysine with peanut meal, which is severely deficient in lysine for channel catfish, improved growth response in the fish.

There is a general belief that fish do not utilize dietary crystalline amino acids as well as chickens or swine, or at least not with conventional once-per-day fish feeding practices. Young carp fed once daily on a diet containing a high level of crystalline amino acids excreted up to 40% of the free amino acids intact through the gills and kidneys (Murai 1985). Increasing the feeding frequency to four times daily improved utilization of the crystalline amino acids. This supports the concept that fish, like swine, do not utilize supplemental crystalline amino acids well when fed once per day because the crystalline amino acids are not absorbed from the gut at the same time as amino acids from the ingested protein.

Protein Requirements of Fish

When feeding guides recommend a minimum level of protein for a fish feed, it should be assumed that the protein is balanced in the essential amino acids. Reports in the technical literature have indicated that the optimum level of protein in feeds for growth of fish has ranged from 25% to 50%. In all of these studies the researchers were probably justified in making their conclusions that a specific percentage of protein was optimum under their experimental conditions because a number of factors influence the growth response of fish to feeds containing different levels of protein. Some of these are size of fish, water temperature, feed allowance, amount of nonprotein energy in the feed, quality of the protein, and natural food availability.

Fish have higher protein requirements during early growth than during later phases of growth. Mangalik (1986) showed that 3-g channel catfish required almost 4 times more protein per day than 250-g fish for maximum growth, but ratio of protein to energy in the diet did not change much (Table 2.2, page 15). He demonstrated that the smaller channel catfish could grow as well from a 27% protein diet as from a 38% protein diet when the energy level was low, but when the energy level increased, diet consumption decreased and the low protein diet would not support maximum growth.

Coldwater and warmwater fish have been shown to respond to higher protein levels at higher water temperatures in both laboratory and commercial culture conditions. This may be because nonprotein energy sources are not digested or metabolized as well as protein at lower temperatures and are thus less efficient in sparing protein.

Natural pond food consumed by fish can be an important protein source. This is influenced by natural pond productivity, feeding behavior of the fish, and density of fish in the pond. Pond sources of protein are primarily of animal origin—high in quality and containing at least 50% protein on a moisture-free basis. Thus, a significant dietary contribution from this source would reduce the protein requirement of the supplemental diet. For example, tilapia and shrimp grow as well on relatively low protein diets (25% or less of protein) as on higher protein diets when natural pond food is a significant part of their diet, but they respond to higher protein diets in an environment with limited natural food.

When fish are underfed, they usually respond to higher amounts of dietary protein. This has been demonstrated in a series of pond feeding experiments with channel catfish. Reasons for interaction between feeding rate and dietary protein percentage for maximum growth are not completely clear. It is reasonable that a high protein diet will come closer than a low protein diet to supplying the fish's protein need for growth during restricted feeding. Also, when fish are underfed, a higher percentage of the dietary protein will be used to meet the metabolic energy needs of the fish unless the energy/protein ratio of the diet is increased.

Protein Synthesis in Fish

Protein synthesis in animal tissue is a complex process which involves deoxyribonucleic acid (DNA), ribonucleic acid (RNA), and ribosomes. DNA, a chromosomal component of cells, carries the genetic information in the cell and transmits inherited characteristics from one generation to the next. The structure of DNA consists of four types of nitrogenous bases (adenine, guanine, cytosine, and thymine) linked to a backbone of alternating phosphate and deoxyribose groups. The DNA molecule is in the form of a long double helix, which consists of two chains running in opposite directions and linked together through hydrogen bonds where adenine always pairs with thymine and guanine pairs with cytosine. The sequence of the nitrogenous bases can vary infinitely and this sequence determines the exact protein to be synthesized. DNA controls the development of the organism by controlling the formation of RNA. The composition of RNA is similar to that of DNA, except that ribose is the sugar instead of deoxyribose and uracil replaces thymine. The nucleotides of RNA are linked through their phosphate groups to form long single chains which might fold and form areas of double helical structure.

There are three kinds of RNA in cells that participate in protein synthesis: messenger RNA, transfer RNA, and ribosomal RNA. Messenger RNA carries the code transcribed from DNA and determines the sequence of amino acids in the protein being formed. Transfer RNA's carry specific amino acids to the ribosomes where they interact with messenger RNA. Ribosomal RNA is part of the structure of the ribosome, which is the site of protein formation in the cell.

Thus, amino acids are linked in sequence predetermined by the sequence of nitrogenous bases in messenger RNA and, in turn, in DNA. The addition of amino acids to a growing polypeptide chain occurs very rapidly. For example, synthesis of the protein chains in hemoglobin occurs at the rate of two amino acid additions to the chain per second so that these proteins, which contain 141 to 146 amino acid residues, are synthesized in about 1.5 minutes.

VITAMINS

Vitamins are organic compounds required in the diet in relatively small quantities for growth, health, and function in animals. A vitamin that is a dietary essential for some animals may not be for other species. For example, humans and other primates, guinea pigs, and most fishes require vitamin C in the diet, but most land animals do not. Although the requirements are small, deficiencies of these micronutrients can cause symptoms ranging from poor appetite to severe tissue deformities. Deficiency signs identified in channel catfish, salmonids, common carp, and Japanese eel are presented in Table 2.5 (pages 30–32).

Vitamins are classified as water soluble and fat soluble. Eight of the water soluble vitamins are required in relatively small quantities and have primarily coenzyme functions; they are known as the water-soluble B complex. Three water-soluble vitamins that are required in larger quantities have functions other than coenzymes and are sometimes referred to as the macrovitamins. This group includes vitamin C, myo-inositol, and choline. Vitamins A, D, E, and K are the fat-soluble vitamins.

Table 2.5. VITAMIN DEFICIENCY SIGNS IN FISH

Vitamin	Channel catfish	Trout and salmon	Common carp	Eel

A	Exophthalmos
	Edema
	Ascites
		Exophthalmos
		Eye lens
		  displacement
		Corneal and
		  retinal
		  degeneration
		Ascites
		Skin
		  depigmentation
			Depigmentation
			Twisted opercula
			Fin and skin
			  hemorrhages
				—[1]

D	Low bone ash	Impaired calcium
		  homeostasis
		Tetany of white
		  skeletal muscle
			—[1]	—

E	Muscular
	  dystrophy
	Exudative
	  diathesis
	Skin
	  depigmentation
	Low hematocrit
	Ceroidosis and
	  hemosiderosis in
	  visceral organs
	Lipid peroxidation
	  of liver
	  microsomes
	Erythrocyte
	  hemolysis
		Anemia
		Variable sized,
		  fragmented
		  erythrocytes
		Ascites
		Muscular
		  dystrophy
		Lipid peroxidation
		  of liver
		  microsomes
		Reduced immune
		  responses
		Skin
		  depigmentation
			Muscular
			  dystrophy
			Exophthalmos
			Kidney, pancreas
			  degeneration
			Ceroids in visceral
			  organs
				Skin and fin
				  hemorrages
				Dermatitis

K	Skin hemorrhages
	Prolonged blood
	  clotting time
		Prolonged blood
		  clotting
		Anemia
			—	—

Thiamin	Dark skin color
	Loss of
	  equilibrium
	Hypersensitivity
	Convulsions
		Hyper-irritability
		Convulsions
		Loss of
		  equilibrium
		Low transketolase
			Fin congestion
			Nervous disorders
			Depigmentation
			Subcutaneous
			  hemorrhage
				Trunk-winding
				Subcutaneous
				  hemorrhage
				Congested fins

Riboflavin	Poor growth
	Short, stubby body
		Lens cataract
		Adehesion of lens
		  and cornea
		Reduced activity of
		  erythrocyte
		  glutathione
		  reductase
		Dark pigmentation
			Skin and fin
			  hemorrhages
			Hemorrhagic heart
			  muscle
			Anterior kidney
			  necrosis
				Dermatitis
				Photophobia
				Fin hemorrhage
				Abdominal
				  hemorrhage

Pyridoxine	Nervous disorders
	Tetany
	Greenish-blue
	  coloration
		Convulsions
		Hyperirritability
		Erratic, spiral
		  swimming
		Rapid breathing,
		  gasping
		Rapid onset of
		  rigor mortis
			Nervous disorders
			Hemorrhage
			Edema
			Low
			  hepatopancreatic
			  transferases
			Dermatitis
				Convulsions
				Nervous disorders

Table 2.5. (*continued*)

Vitamin	Channel catfish	Trout and salmon	Common carp	Eel
Pantothenic acid	Clubbed, exudate-covered gills Anemia Erosion of skin, barbels, lower jaw	Clubbed, exudate-covered gills Anemia Atrophied pancreatic acinar cells Vacuoles and hyaline bodies in kidney tubules	Anemia Exophthalmos	Dermatitis Congested skin Hemorrhagic skin Abnormal swimming
Biotin	Skin depigmentation Anemia Reduced liver pyruvate carboxylase	Degeneration of gill lamellae Skin lesions Fatty liver Reduced acetyl CoA and pyruvate carboxylase Altered fatty acid synthesis Degeneration of pancreatic acinar cells	Poor growth Increased dermal mucous cells	Abnormal swimming Dark skin
Niacin	Skin and fin lesions Exophthalmos Anemia Deformed jaws	Skin and fin lesions Colon lesions Anemia Photosensitivity Sunburn	Skin hemorrhages lesions	Poor swimming coordination Dark color Skin lesions Anemia
Folic acid	Reduced growth	Anemia Pale gills Large, segmented erythrocytes	No deficiency found	Poor growth Dark coloration
B_{12}	Reduced hematocrit	Anemia Small, fragmented erythrocytes	No deficiency found	Poor growth
C	Lordosis Scoliosis Reduced bone collagen Increased sensitivity to bacterial infection Slow wound repair Reduced hematocrit	Lordosis Scoliosis Hemorrhagic exophthalmos Ascites Reduced hematocrit Deformed operculum Abnormal histology of support cartilage in eye and gill	No deficiency found	Fin and dermal hemorrhages Lower jaw erosion
Choline	Fatty liver Hemorrhagic kidney and intestine	Fatty liver	Fatty liver	White-grey intestine

Table 2.5. *(continued)*

Vitamin	Channel catfish	Trout and salmon	Common carp	Eel
Inositol	No deficiency found	Slow gastric emptying Reduced cholinesterase and transaminase Fatty liver Decreased phosphatides	Skin lesions	White-grey intestine

Note: Anorexia and reduced growth rate are generally characteristic of vitamin deficiencies in fish. These signs are not included in the table unless they are the only ones observed. Mortality is also excluded unless it is an immediate and primary sign.
[1] — means not evaluated.

Essentiality of all of the 15 vitamins has been established for fish, although all fish species do not seem to have a dietary requirement for all 15 of the vitamins. Qualitative and quantitative needs for vitamins have been studied with several species using the controlled environment, purified diet approach. Vitamin requirements for fish vary with species, size, growth rate, nutrient interrelationships, environment (temperature, toxicants), and metabolic function (growth, stress response, disease resistance). Intestinal microorganism synthesis is a significant source of vitamins for some species. Culture system and feeding habits of the fish influence the need for vitamin supplementation of practical fish feeds; fish feeding actively on natural aquatic organisms may not need certain vitamins in the supplemental diet.

Vitamin requirements for channel catfish, common carp, and salmonids are presented in Table 2.6. These values represent minimum requirements for growth of young fish, determined under near-optimum growing conditions, with no allowance for processing or storage losses. Therefore, the vitamin levels presented in the table should be increased by 25% to 100%, depending upon the vitamin, to allow for losses in feed processing and storage, and possible increased needs by the fish due to stress, infection, or interaction with other substances in the feed.

Vitamin A

Vitamin A is found only in the animal kingdom. It exists in the free alcohol form as retinol and as esters of higher fatty acids. One International Unit (I.U.) of vitamin A is equal to 0.3 μg of all-trans retinol. Vitamin A_1, $C_{20}H_{30}O$, has been isolated from lipids of many

Table 2.6. MINIMUM VITAMIN REQUIREMENTS FOR GROWTH OF YOUNG FISH
(AMOUNT PER KG OF DIET)

Vitamin	Units	Channel catfish[1]	Common carp[1]	Salmonids[2]
A	I.U.	1,000–2,000	R	2,500
D	I.U.	500–1,000	—	2,400
E	I.U.	50	R	30
K	mg	R[3]	—	10
Thiamin	mg	1	1	10
Riboflavin	mg	9	8	20
Pyridoxine	mg	3	6	10
Pantothenic acid	mg	20	30–50	40
Niacin	mg	14	28	150
Folic acid	mg	R	N	5
B$_{12}$	mg	R	N	0.02
Biotin	mg	R	R	0.1
Inositol	mg	N[3]	10	400
Choline	mg	R	4,000	3,000
Vitamin C	mg	60	R	100

[1] From NRC (1983), except vitamin E requirement for channel catfish, which is from Wilson, Bowser, and Poe (1984).
[2] From NRC (1981), except for biotin requirement for salmonids, which is from Castledine et al. (1978).
[3] R means essential but dietary requirement not determined. N means no dietary requirement found under defined experimental conditions.

fishes and land animals, whereas vitamin A$_2$, C$_{20}$H$_{28}$O, has been isolated only from freshwater fishes (Figure 2.14).

Plants produce red-to-yellow pigmented compounds called carotenoids, some of which have vitamin A activity. Beta-carotene (Figure 2.15) has by far the highest vitamin A activity. This compound appears to be capable of yielding two moles of retinol upon simple hydrolytic cleavage; however, the vitamin A activity of beta-carotene is much less than this for most animals. Several fishes have been found capable of using beta-carotene for vitamin A activity; however, Poston et al.

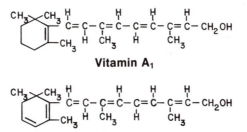

Fig. 2.14. Vitamin A.

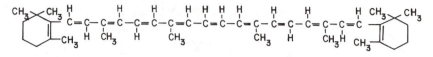

Fig. 2.15. Beta-carotene.

(1977) found that coldwater fish could utilize beta-carotene at 14° C, but not at 9° C. Lee (1987) found that channel catfish readily converted beta-carotene into vitamins A_1 and A_2 in almost a 1 to 1 ratio.

Vitamin A, like other fat soluble vitamins, is stored in large amounts in the body (liver) if intake exceeds metabolic need. It is possible for fish to store enough vitamin A to produce a toxic condition (hypervitaminosis); however, prolonged consumption of a diet with an unusually high level of vitamin A would be required to produce this.

An established physiological function of vitamin A is its role in vision. Retinol is combined with a protein, opsin, to form rhodopsin, which is the compound involved in the photochemical reaction in the retina of the eye in the process of vision.

Other metabolic roles of vitamin A are less well understood. It is used for maintenance of mucosal membranes that line many body organs, the gastrointestinal tract, the respiratory tract, and the eye. In vitamin A deficiency, epithelial cells fail to differentiate beyond the squamous type to mucus-secreting type, and mesenchymal cells fail to differentiate beyond the blast stage. Epithelial cells from the eye and many other areas of the body keratinize and this lowers resistance to infection. Several deficiency symptoms appear to be related to impaired function of epithelial tissue. Reproduction is impaired in most animals.

Vitamin A deficiency in salmonids has been described as reduced growth rate, light skin color, ascites (fluid in abdominal cavity), and pathological condition of the eye characterized by exophthalmos, hemorrhagic eyes, eye lense displacement, thinning of cornea, degradation of the retina, and twisted gill opercula (see Figure 2.16). Channel catfish fed vitamin A deficient diets over a long period (2 years) developed exophthalmos, edema (collection of fluid in tissues) and kidney hemorrhage. Common carp showed deficiency signs of light skin color, fin and skin hemorrhages, exophthalmos, and deformed gill opercula.

Retinol and beta-carotene are sensitive to oxidation, so natural sources of vitamin A may be oxidized to various degrees. Thus, supplemental vitamin A should be added to fish feeds. Synthetic vitamin A used in fish feeds is in stabilized forms, usually as palmi-

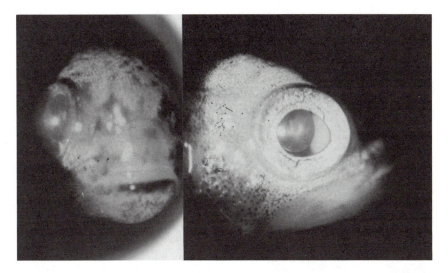

Fig. 2.16. Vitamin A deficiency in fish often causes eye disorders. Brook trout with corneal edema and enlarged, misshapened eye globes (left) and displaced lens (right).
(Courtesy of H. A. Poston)

tate, acetate, or propionate esters. Dry additives are usually in beadlet form, where the retinol ester is coated with gelatin or some other oxygen barrier.

Vitamin E

Vitamin E is present in at least eight tocopherols which occur in plants. The name *tocopherol* is from the Greek word *tocos,* which means childbirth. Alpha-tocopherol (Figure 2.17) has the highest vitamin E activity. Vitamin E activity of compounds is measured in International Units (I.U.), with 1 I.U. being equivalent to the biological activity of 1 mg of D-α-tocopherol (replacing the formerly used

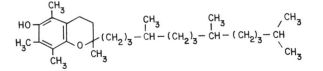

Fig. 2.17. Alpha-tocopherol.

DL-α-tocopherol acetate). Traditionally, biological activity has been based on prevention of fetal resorption in rats.

Commercial antioxidants, like ethoxyquin, which have no chemical relationship to the tocopherols, also have vitamin E activity in fish and other animals. Lovell, Miyazaki, and Rebegnator (1984) found that 125 mg of ethoxyquin/kg of diet prevented deficiency signs in channel catfish, which included nutritional muscular dystrophy, but was not as effective as 25 mg of alpha-tocopherol/kg of diet for growth of the fish. Thus, under certain economic conditions, commercial antioxidants may spare some of the alpha-tocopherol in fish feeds.

A major function of vitamin E is its role as a metabolic antioxidant, with a specific role in preventing oxidation of unsaturated phospholipids in cellular membranes, such as erythrocytes, and subcellular membranes such as mitochondria. It is often referred to as a metabolic free radical scavenger or peroxide scavenger. In most animal species, an increase in polyunsaturated fatty acids, especially when partially oxidized, in the diet produces an increase in dietary need for vitamin E. The function of vitamin E as an antioxidant is evidenced by an increased need in the absence of selenium. Selenium is a component of the enzyme glutathione peroxidase, which catalyzes the removal of metabolic peroxides. A specific role for vitamin E in an enzyme system has not been identified, although impairment of several enzyme systems, such as those involved in porphyrin and heme synthesis, has been identified in vitamin E deficiencies in various animals.

Diverse physiologic abnormalities have been demonstrated with vitamin E deprivation in animals. Common are nutritional muscular dystrophy, which has been produced in several fishes and terrestrial animals (see Figure 2.18); and pathological conditions of male and female reproductive organs, causing reduced fertility and reproduction. Increased permeability of capillaries, which result in hemorrhages and edema in various body areas, has been caused by vitamin E deficiency in various animals. The syndrome exudative diathesis, manifested by accumulations of fluid under the skin or in the abdominal cavity (sometimes of a greenish color caused by decomposed hemoglobin), has been produced in channel catfish and salmonids fed vitamin E–deficient diets. Vitamin E deficiency causes reduction in ability of erythrocytes to withstand peroxide deterioration of membranes. Severe anemia, characterized by immature, irregularly shaped and sized erythrocytes, is produced in vitamin E deficiency.

Other vitamin E deficiency signs described for several fish species include fatty livers and ceroid (dark lipoid) bodies in liver. "Sekoke disease" in common carp, characterized by thinning of flesh on the

Fig. 2.18. Channel catfish at right was fed a vitamin E deficient diet and shows nutritional muscular dystrophy. The other fish were fed diets containing 125 mg/kg of ethoxyquin (left) or 25 mg/kg of α-tocopherol (center) and showed no myopathy.

back of the fish, was caused by feeding oxidized silkworm pupae, but corrected by supplementing the diet with vitamin E.

Subclinical measurements used to detect vitamin E deficiency include erythrocyte fragility and histological examination of tissues for necrosis of muscle fibers and ceroid concentration in liver and kidney.

In most experimental and practical situations where the classical vitamin E deficiency syndrome in fish has been produced, inclusion of high levels of polyunsaturated fatty acids or omission of selenium from the diet has been necessary. However, Lovell, Miyazaki, and Rebegnator (1984) fed channel catfish diets low in polyunsaturated fatty acids, using stripped lard as the lipid source, and produced fish with severe nutritional muscular dystrophy and other signs of vitamin E deficiency. This implies that fish feeds not containing high levels of polyunsaturated lipids, such as many commercial warmwater fish feeds, can cause reduced growth rate and various pathologies when deficient in vitamin E.

Several studies have found rainbow trout to be relatively insensitive to vitamin E deficiency unless oxidized polyunsaturated fats were included in the diet. Cowey et al. (1984) reduced the water temperature from 15° C to 6° C and produced severe myopathy in rainbow trout. This indicated that maintenance of fluidity in biomembranes is more demanding in fish at lower temperatures.

Hypervitaminosis E can be produced in fish. Rainbow trout fed about 100 times the requirement, 5,000 mg of DL-α-tocopherol/kg of diet, showed reduced blood concentration of erythrocytes.

Alpha-tocopherol is found in most plant seeds. Significant sources are plant oils, germ or gluten meals, distillery or brewery dried products, and grain brans and by-products. Whole grains, fish meal, and solvent-extracted oilseed meals are poor to fair sources and need supplementation for feeding fish.

Vitamin D

Vitamin D is found in nature in two forms: ergocalciferol (D_2) and cholecalciferol (D_3) (Figure 2.19). Ultraviolet irradiation of two provitamins, ergosterol (found in plants) and 7-dehydrocholesterol (found in skin of animals), will produce vitamins D_2 and D_3, respectively. Animals not exposed to sunlight need a dietary source of vitamin D. Fish get relatively little ultraviolet energy from the sun because of the shallow depth of penetration of these rays in natural waters. Most land animals, except chickens, can use D_2 and D_3 interchangeably. Fish, however, utilize D_2 poorly or not at all. Rainbow trout used D_3 about

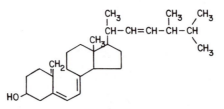

Vitamin D$_2$

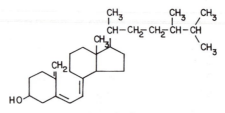

Vitamin D$_3$

Fig. 2.19.   Two forms of vitamin D.

three times more efficiently than D_2. One I.U. of vitamin D activity is equivalent to the antirachitic effect of 0.025 μg of cholecalciferol.

At least two fish species have shown a dietary need for vitamin D_3. Although little research has been done on vitamin D metabolism in fish, it can be assumed that a physiological role of vitamin D_3 in fish is similar to that in warmblooded animals, where it is the precursor to 1, 25-dihydroxycholecalciferol, a major calcium-regulating hormone and an important phosphate-regulating hormone. Vitamin D, after absorption from the intestine, is converted in the liver to 25-hydroxycholecalciferol and subsequently in the kidney to 1, 25-dihydroxycholecalciferol, the active hormone. This hormone is responsible for maintaining serum calcium and phosphorus levels by altering rate of intestinal absorption, renal resorption, and bone mobilization. It is also thought to play a role in the synthesis of calcium and phosphorus transport proteins.

Channel catfish fed a vitamin D–deficient diet for 16 weeks showed reduced weight gain and decreased body levels of ash, calcium, and phosphorus. Signs of vitamin D deficiency in rainbow trout have been described as slow growth rate, tetany of white muscle, ultrastructural changes in epaxial white muscle fibers, and elevated levels of triiodothyronine in blood.

Fish oil is usually a good source of vitamin D_3 but other animal products are generally poor. Plant products are devoid of D_3. Because of the scarcity of vitamin D in feedstuffs, this vitamin should be supplemented in commercial fish feeds.

Vitamin K

The name *vitamin K* was given to a fat-soluble factor necessary in the diet of chicks to prevent hemorrhage and for normal blood clotting. Several compounds with vitamin K activity have been isolated or synthesized. These include phylloquinone (vitamin K_1), found in green plant leaves; and menaquinone (vitamin K_2), isolated from fishmeal and animal feces. Menadione (K_3) is a synthesized compound and has more vitamin K activity that K_1 or K_2 (Figure 2.20).

Fig. 2.20.   Vitamin K_3 (menadione).

Vitamin K is necessary for normal blood clotting in all animals, including fish. Several proteins necessary for blood coagulation are dependent upon vitamin K for their synthesis. These include prothrombin, proconvertin, plasma thromboplastin antecedent, and Stewart-Prower factor. The role of vitamin K is assumed to be in conversion of the precursor to the factor (prothromin to thrombin) by the carboxylation of the glutamic acid residues to form γ-carboxyglutamic acid in the active proteins. Other proteins, such as in bone and kidney, have been found that contain γ-carboxyglutamyl acid residues, so it is presumed that vitamin K is involved in other carboxylase enzyme systems besides blood clotting.

Channel catfish and trout require dietary vitamin K for normal blood coagulation. Growth rate was not affected in either of the fishes when vitamin K was deleted from the diets. Quantitative requirement for vitamin K has not been determined for warmwater fishes. Levels of 0.5 mg to 1 mg of menadione per kg of diet is sufficient to maintain normal blood coagulation in fingerling trout.

Intestinal synthesis is an important source of vitamin K in some animals. This source has not been evaluated in fish, but dietary sulfaguanidine and low water temperature each prolonged blood coagulation time in trout.

Fish meal and alfalfa meal are good sources of vitamin K. Menadione sodium bisulfite or menadione dimethylpyrimidinal bisulfite are synthetic sources used in commercial feeds; the latter is more heat stable during feed processing.

Thiamin

Thiamin is found in many grains and seeds, being most concentrated in the seed coat. Discovery of this growth factor began when persons consuming polished rice developed beriberi, but recovered when a factor in the seed coat was added to the diet. Thiamin is composed of a pyrimidine ring and a thiazole ring, as shown in Figure 2.21. It is synthesized by higher plants, but not by animals. It is relatively

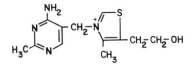

Fig. 2.21. Thiamin.

sensitive to heat and moisture at pH above 5, so the commercial vitamin is sold in the hydrochloride form to ensure stability.

The active form of the vitamin is thiamin pyrophosphate. Phosphorylation occurs in the liver. Thiamin pyrophosphate acts as a coenzyme for several metabolic decarboxylation and transketolation reactions. It is involved in decarboxylation of pyruvic acid and α-ketoglutaric acid in aerobic glycolysis. It also acts as a coenzyme in transketolation in metabolism of glucose through the pentose phosphate shunt. Transketolase activity in erythrocytes and kidney is a sensitive indicator for thiamin status in rainbow trout.

Thiamin deficiency affects the central nervous system in fish, birds, and mammals. Polyneuritis in chicks (lack of control of position of the head) and Chastek paralysis in mink and foxes are caused by thiamin deficiency. Thiamin deficiency in fish causes hypersensitivity to disturbance, loss of equilibrium, and convulsions. Fish show thiamin deficiency quickly: channel catfish in 6 to 8 weeks, carp in 8 weeks, and Japanese eel in 10 weeks.

Tissues of most fishes contain thiaminase, an enzyme that can destroy thiamin in nonliving tissue by splitting it into its two component ring structures. Heating the fish destroys the enzyme. Feeding fish or fish visceral organs without heat treatment has produced thiamin deficiency problems in mink and foxes and in channel catfish. The thiamin is destroyed prior to ingestion, when the enzyme (thiaminase) and substrate (thiamin) are in contact for a period of time. Channel catfish fed diets containing 40% nonheated fish viscera developed thiamin deficiency in 10 weeks, but when the fish were fed an additional diet containing thiamin in a separate meal daily, no deficiency occurred.

Riboflavin

Riboflavin is widely found in nature; green plants, seed coat of grains, and yeast are rich sources. Its origin is plant or microorganism synthesis. Nonruminant animals require it in their diet. Chemical structure of riboflavin (Figure 2.22) consists of a dimethyl-isoalloxazine moiety conjugated with ribose through which the vitamin is linked to phosphate in the intestinal wall to form the active coenzyme.

Riboflavin is a component of two flavoprotein coenzymes, flavin mononucleotide (FMN) and flavine adenine dinucleotide (FAD), which are components of prosthetic groups of oxidases and reductases that act upon metabolic degradation products of proteins, carbohydrates, and lipids.

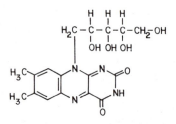

Fig. 2.22.   Riboflavin.

A number of riboflavin deficiency signs have been identified in animals, but none have been related to a specific biochemical role of the vitamin. Crooked and stiff legs occur in swine and chickens (curled-toe paralysis). Eye problems occur in humans, farm animals, and fish. Photophobia and cataracts have been found in several fish deprived of riboflavin (see Figure 2.23). Dark skin occurred in salmonids; short, stubby bodies were described in riboflavin-deprived channel catfish; dermatitis and hemorrhagic fins were found in eel; and necrosis of head kidney was observed in carp.

A sensitive subclinical test for riboflavin deficiency is in vitro measurement of erythrocyte glutathione reductase (EGR) activity in presence of added FAD. Approximately 3 mg of riboflavin per kg of diet is sufficient to prevent change in EGR activity in rainbow trout. The minimum dietary requirement for normal growth and to prevent dwarfism in channel catfish is 9 mg/kg of body weight.

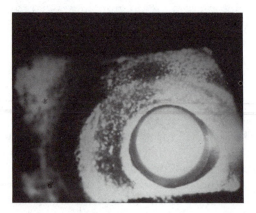

Fig. 2.23.   Severe eye cataract in rainbow trout fed riboflavin deficient diet (above). Normal lense (upper right).
(Courtesy of Tunison Laboratory of Fish Nutrition)

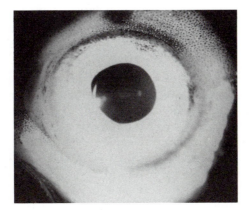

Fig. 2.23. (*continued*)

Most feedstuffs are reasonably good sources of riboflavin, except whole grains. Bran or polishings from grains and distillery byproducts are fair to good sources.

Niacin

Niacin, nicotinic acid, and nicotinamide are often used interchangeably. Niacin is actually the generic descriptor for pyridine 3-carboxylic acid and derivatives exhibiting biological activity of nicotinamide (Figure 2.24). Nicotinamide occurs in physiological systems as a component of two coenzymes of the hydrogen transport system, nicotinamide adenine dinucleotide (NAD) and nicotinamide adenine dinucleotide phosphate (NADP). These coenzymes are involved in a number of oxidation–reduction reactions. NAD is specific for hydrogenases involved in passing electrons on to oxygen in the electron transport systems. NADP is specific in hydrogenases such as in fatty acid synthesis and the pentase phosphate shunt in glucose metabolism.

Niacin deficiency causes pellagra in humans, which is characterized primarily by dermatological problems, and is commonly found in situations where the diet is limited to corn or polished rice. Most fish show niacin deficiency signs rather quickly. Skin lesions are a common

Fig. 2.24. Niacin.

deficiency sign in fish. Channel catfish show skin and fin lesions along with deformed jaws, exophthalmia, anemia, and high mortality rate. Eels show skin lesions, dark pigmentation, and ataxia. Salmonids show sensitivity of skin to sunburn along with fin erosion, intestinal lesions, and muscle weakness. Exposure of rainbow trout to ultraviolet irradiation enhanced the effects of niacin deficiency and increased niacin requirement (see Figure 2.25).

Niacin occurs naturally as an amide in all living tissue. Although plant seeds contain substantial amounts of niacin, the niacin occurs mostly in bound form and is highly unavailable to animals. Most land animals can convert the amino acid tryptophan to niacin and this source can contribute to their niacin requirement. However, fish appear to have limited ability to make this conversion. Brook trout were found incapable of converting tryptophan to niacin efficiently. Also, niacin deficiency can be demonstrated rapidly in fishes fed

Fig. 2.25.   Niacin deficiency in rainbow trout. Top fish is control; two lower fish were fed a niacin-deficient diet for 16 wk. Bottom fish was exposed to 20 hr/day of ultraviolet irradiation and had more serious fin erosion.
(Courtesy of S. G. Hughes)

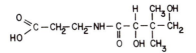

Fig. 2.26. Pantothenic acid.

tryptophan in the diet. Because of the presumed limited availability of niacin in grains and oil seed meals to fish, vitamin supplementation of practical fish feeds is recommended.

Pantothenic Acid

Pantothenic acid is synthesized by plants and microorganisms. Chemically, it is composed of pantoic acid (2,4-dihydroxy-3,3-dimethyl butyric acid) linked through a peptide bond to β-alanine (Figure 2.26). In nature, it occurs largely in bound form as coenzyme A. Because of its sensitivity to destruction by heat and high or low pH, it is available commercially as calcium or sodium salt.

The only known function of pantothenic acid is as a component of coenzyme A (CoA), which is involved in transfer of acetyl (2-carbon) units in numerous reactions in the metabolism of proteins, carbohydrates, and lipids. Coenzyme A contains pantothenic acid linked through pyrophosphate to adenosine 3'-phosphate on one side and β-mercaptoethylamine on the other (Figure 2.27). The acceptance of the acetyl units, which help form acetyl CoA, is through the sulfhydryl group of β-mercaptomethylamine. Coenzyme A formation is an important route for carbohydrates and fatty acids to enter the tricarboxylic

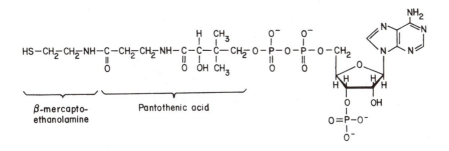

β-mercapto-
ethanolamine

Pantothenic acid

Adenosine,3'- mono- 5'-diphosphate

Fig. 2.27. Coenzyme A (CoA).

acid cycle. Coenzyme A has manifold functions, anabolic and catabolic, involving cellular energy release and synthesis of steroids, sphingosine, porphyrin, and many other compounds.

Pantothenic acid is a dietary essential for birds, nonruminant mammals, and fish. A dietary deficiency results in growth failure in all. Skin and hair problems are common deficiency characteristics. Pantothenic acid deficiency results in impaired functions of mitochondrial-rich, high energy expenditure cells. This may partially explain why gill lamellae, kidney tubules, and acinar cells in pancrease, all of which carry on a high level of metabolic activity, are all sensitive to pantothenic acid deficiency.

Fish show pantothenic acid deficiency signs quickly. Channel catfish fry showed deficiency signs within 2 weeks and subadult fish became anorexic in 4 weeks. Deficiency causes dietary gill disease in several fish species, which is characterized by clubbed, exudate-covered gill lamellae, fused gill filaments, and swollen operculums (see Figure 2.28). Lesions on skin and fins along with anorexia and poor growth are also found in fish deprived of pantothenic acid.

Fig. 2.28. Fused gill lamellae and filaments of rainbow trout fed a pantothenic acid-deficient diet (left) and normal gill filament (right). *(Courtesy of C. Y. Cho and W. D. Woodward)*

Pantothenic acid is widely found in commercial feedstuffs, but the level and availability in processed feed are likely to be lower than the requirement for most fishes. Supplemental pantothenic acid is recommended for commercial fish feeds, especially for small fish.

B_6 (Pyridoxine)

Three chemically related compounds that have similar metabolic functions have been identified: pyridoxine, pyridoxamine, and pyridoxal. Pyridoxine (Figure 2.29) was the first identified and was given this name. Later, the other two were identified and the name B_6 was given to this group of compounds. Vitamin B_6 is now the approved name for this vitamin. Vitamin B_6 is a dietary requirement of all nonruminant mammals, birds, and fish. Vitamin B_6 compounds are synthesized by plants and some microorganisms. It is widely found in feedstuffs of plant and animal origin.

The metabolically active B_6 coenzyme is pyridoxal phosphate. It is functional in a number of enzymes in which amino acids are metabolized, including decarboxylases, transaminases, sulfhydrases, and hydroxylases. Increasing protein percentage in the diet causes an increase in the B_6 requirement of salmonids. Because fish are fed much higher protein diets than land animals, it should be expected that the B_6 requirement would be higher for fish. Vitamin B_6 is involved in metabolism of carbohydrates and lipids. It is essential for the synthesis of heme (in hemoglobin), and in the synthesis of serotonin from tryptophan, which may explain why B_6 deficiency causes nervous disorders.

Deficiency signs in fish develop quickly. Young channel catfish, salmonids, and carp show deficiency signs in 4 to 8 weeks. These include nervous disorders, such as hypersensitivity to disturbances; poor swimming coordination; convulsions; and tetany when handled. Channel catfish develop a greenish-blue sheen to the skin. Common carp show edema, exophthalmos, and skin lesions.

The requirement for fish is 3 mg/kg to 10 mg/kg of diet, and most commercial feedstuffs contain this amount. However, because of varia-

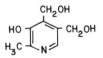

Fig. 2.29.   Vitamin B_6 (pyridoxine).

tion and uncertainty of content in feedstuffs and in processed, stored feeds, vitamin B_6 supplementation in commercial fish feeds is practiced.

Biotin

Biotin (Figure 2.30) was first recognized in its association to "egg-white injury" in rats because typical deficiency signs resulted when raw egg white was supplemented into most diets. When liver was fed, egg white injury was prevented. The preventive and causative factors were subsequently identified as biotin and avidin, the latter a heat labile protein in egg white that combines with biotin and renders it inactive.

Biotin functions as a coenzyme in carboxylation reactions. Enzyme systems containing biotin are acetyl-CoA carboxylase, propionyl-CoA carboxylase, and pyruvate carboxylase. Metabolic functions requiring biotin are many and include synthesis of fatty acids, oxidation of energy yielding compounds, synthesis of purines, and deamination of certain amino acids. A reliable indicator of biotin status that has been used in fish and warmblooded animals is pyruvate carboxylase or acetyl-CoA carboxylase activity.

Biotin deficiency is difficult to produce in nonruminant mammals without feeding raw egg white. Deficiency signs in chicks and several fishes have been produced, however, without feeding a biotin antagonist. Fish vary in sensitivity to biotin deficiency. Young channel catfish required 14 weeks, while young rainbow trout required only 4 weeks to show deficiency signs, each under optimum environment for growth. Channel catfish showed reduced growth rate, light skin color, and hypersensitivity to abrupt noise or movement. Rainbow trout showed poor growth, degeneration of gill lamellae and epithelium, and enlarged, pale livers. Subclinical signs of fish are reduced carboxylase activities in liver and, in trout, abnormal synthesis of glycogen and fatty acids, degeneration of acinar cells of pancrease, and glycogen deposition in kidney tubules.

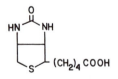

Fig. 2.30. Biotin.

Bioavailability of biotin in many feedstuffs is limited. For example, one-half or less of the total biotin in wheat, barley, sorghum, meat and bone meal, and fish meal is available to chickens. Biotin availability in corn and soybean meal, however, is higher. Channel catfish do not require supplemental biotin in practical corn-soybean meal or corn-soybean meal-fishmeal diets for normal growth and pyruvate carboxylase activity. Rainbow trout also do not benefit from biotin supplementation of practical diets. The dietary requirement for fish is low, about 0.25 mg/kg of diet, so practical fish diets do not usually need supplemental biotin.

Folic Acid

When chemically identified, folic acid was named pteroylglutamic acid. As seen in Figure 2.31, it is composed of glutamic acid (right) plus pteroic acid, the latter being made up of paraminobenzoic acid and a pteridine nucleus. A number of biologically active forms or derivatives of folic acid exist in nature. These include tetrahydrofolic acid, which is the active coenzyme form; 5-methyl-tetrahydrafolic acid; folic acid glutamates; and others.

The folic acid coenzyme is functional in the transfer of single-carbon units, a role analogous to that of pantothenic acid in the transfer of two-carbon units. The one-carbon units may be formyl, methyl, hydroxymethyl, or others. Among the reactions involving tetrahydrofolic acid are the synthesis of purines and pyrimidines for formation of nucleic acids.

In animal species where folic acid deficiencies occur, megablastic anemia is usually found, characterized by large, immature erythrocytes. Megablastic anemia has been produced in salmonids, character-

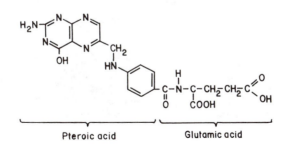

Fig. 2.31. Folic acid.

ized by reduced production of erythrocytes, large and segmented erythrocytes with constricted nuclei, and abnormally large proerythrocytes in the erythropoietic tissues. Folic acid deficiency could not be demonstrated in common carp. Poor growth was seen in folic acid–deficient eel and channel catfish. Combined deficiencies of folic acid and vitamin B_{12} enhance anemia in fish (see Figures 2.32 and 2.33).

In rats and pigs, folic acid deficiency could not be produced without inclusion of an antibiotic in the diet to inhibit intestinal synthesis of the vitamin. Kashiwada, Teshima, and Kanazawa (1970) showed that folic acid is synthesized by intestinal bacteria in common carp; this may explain why the researchers were unable to demonstrate dietary folic acid deficiency.

Butterworth, Plumb, and Grizzle (1986) associated the condition of "no blood" disease in channel catfish with folic acid degradation in the feed. The disease is characterized by very low erythrocyte concentrations in blood, kidneys, liver, and gills, and abnormal erythrocytes characteristic of those in megoblastic anemia. They isolated bacteria from feeds fed to affected fish (there are also molds) that are capable of degrading folic acid to glutamic acid and pteroic acid. Pteroic acid not

Fig. 2.32. Folic acid deficient coho salmon. Note severely pale gills on fish on left and right as compared to normal fish in center.
(Courtesy of Charlie E. Smith)

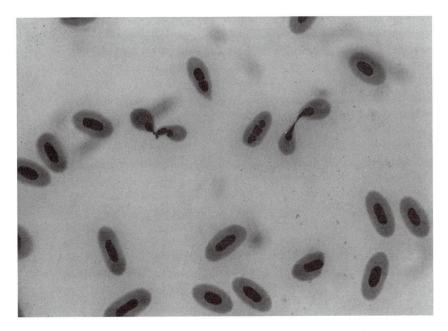

Fig. 2.33. Erythrocytes from folic acid-deficient coho salmon. Note bilobed erythrocytes and segmented nuclei.
(Courtesy of Charlie E. Smith)

only has no folate activity, but is antagonistic to absorption and metabolism of folic acid in warmblooded animals.

Folic acid is synthesized by plants and microorganisms. Grains, oilseed meals, and animal byproducts are all good sources. The needs of farm animals are readily met with practical feeds. Presently, supplemental folic acid is added to most fish feeds. One to five milligrams per kilogram of diet is sufficient for young salmonids.

Vitamin B$_{12}$ (Cyanocobalamin)

This was the last of the 15 recognized vitamins to be identified; it is also the most chemically complex. For many years it was known that liver contained a factor that would correct pernicious anemia. Supplementation of all-plant diets with animal by-products provided a missing growth factor for chickens. In 1948 the factor was isolated from liver and chemical identification was completed in 1955. The molecule (Figure 2.34), which has a molecular weight of 1,354, has a cobalt nucleus in a tetra-ring porphyrin structure. Several similar compounds, where the cyanide or nucleotide is substituted, that have B_{12} activity are found in nature; collectively, they are called cobalamins.

Vitamin B_{12} functions as a coenzyme in a number of metabolic reactions. A specific function is to act in concert with folic acid in the transfer of single-carbon units, such as methylation of uracil to form thymine in DNA synthesis and in methyl transfer in methionine synthesis. Vitamin B_{12} is necessary for normal growth, maturation of erythrocytes, and healthy nervous tissue. A B_{12} deficiency could cause a folic acid deficiency because it is necessary for conversion of tetrahydrofolic acid to its coenzyme form. Intrinsic factor, a mucoprotein in the digestive tract, is necessary for proper absorption of B_{12}. It has been found in the gut of most animals.

Vitamin B_{12} deficiency causes pernicious anemia in humans, characterized by macrocytic anemia and nervous disorders. In other animals

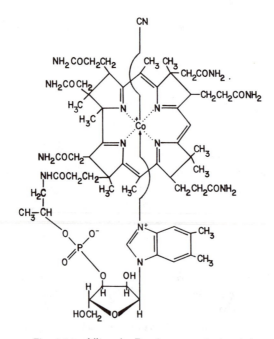

Fig. 2.34.   Vitamin B_{12} (cyanocobalamin).

there is usually microcytic or normocytic anemia and suppressed growth. In fish, salmonids showed microcytic anemia and fragmented erythrocytes with low hemoglobin values. Anemia was not found in B_{12}-deficient channel catfish or carp, but slight reduction in growth occurred in the catfish. Evidence has been presented that intestinal synthesis is a significant source of B_{12} for channel catfish.

Vitamin B_{12} is assumed to be synthesized only by microorganisms. It is a metabolic essential for all animals. Production of a dietary deficiency has been difficult in some monogastric animals; intestinal synthesis was assumed to be the cause.

Inositol

It is widely distributed in plant and animal tissues. In animals it occurs freely as myoinositol, or as a component of phospholipids, as lipositol, in cell membranes. In plants, it is most concentrated in seeds and about two thirds of it is in the form of a hexaphosphate ester, phytic acid, which is highly indigestible to fish and other monogastric animals.

Inositol (Figure 2.35) has no known coenzyme function. Besides being a component of cell membranes, it apparently has some lipotropic action. Rainbow trout fed inositol-deficient diets had large accumulations of triglycerides and cholesterol but low levels of phospholipids in the liver. In several fish species (trout, shrimp, sea bream, carp, and eel), reduced growth, anemia, fin erosion, and slow rate of gut emptying have been reported with inositol deficiency. Inositol deficiency is difficult to produce in some animals when the diet is nutritionally complete otherwise. Burtle (1981) was unable to demonstrate inositol deficiency in channel catfish, and found a significant rate (equal to that in rat) of inositol synthetase activity in the liver.

Inositol   **Phytic acid**

Fig. 2.35.   Inositol and its hexaphosphate ester, phytic acid.

Choline

Choline (Figure 2.36) has no known coenzyme function, but does have several metabolic roles. Its three methyl groups ($-CH_3$) in the molecule make it an important methyl donor. It reacts with acetyl coenzyme A to form acetylcholine, the neurotransmitter. It is a component of lecithin.

Choline can be synthesized in the body if "labile" methyl groups are available. Methionine or cystine may contribute methyl groups to ethanolamine to form choline. Although synthesis in the body is usually not fast enough to meet the choline requirement for normal growth, the dietary content of methionine or cystine as methyl donors or of folic acid and vitamin B_{12} for de novo synthesis of methyl groups can influence the dietary requirement of choline.

Choline deficiency has been produced in most animals with the exception of humans. A common deficiency sign is fatty livers. All fish species evaluated require choline in the diet. In addition to fatty livers, channel catfish had hemorrhagic kidneys and intestines, and eels had white-gray colored intestines.

Choline is widely found in plant seeds. This fact, plus the ability of animals to synthesize it should limit the need to supplement fish feeds with choline. However, feeding fish feeds that contain fat-extracted oilseed as a major ingredient may require choline supplementation because choline is removed with the fat. High methionine diets will reduce the need for dietary choline.

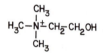

Fig. 2.36. Choline.

Vitamin C (Ascorbic Acid)

Finfish as well as shellfish have been found to be highly sensitive to dietary deficiency of vitamin C, especially the young fish. Gross signs, such as deformed spinal column, distorted gill support cartilage, hemorrhagic areas under the skin, depigmentation, and slow wound healing, have been produced in several fish species by feeding vitamin C-deficient diets.

Most animals can synthesize vitamin C in sufficient quantity for normal growth and function, but a few, such as primates, guinea pigs, some birds, and many fishes cannot because they lack the enzyme L-gulonolactone oxidase for synthesis of vitamin C from glucose. The vitamin occurs in two forms, a reduced form (ascorbic acid) and an oxidized form (dehydroascorbic acid) (Fig. 2.37). The reduced form predominates, but the forms are biologically reversible, so both have vitamin C activity. If the dehydro- form is further oxidized to diketo-gulonic acid, it loses its activity and the reaction is irreversible.

Vitamin C is a strong metabolic reducing agent. Its role in hydroxyl-ation of proline and lysine to the hydroxy-amino acids for the conversion of procollagen to collagen has long been recognized. Many of the deficiency signs in fish are related to malsynthesis of collagen, which is a component of bone, gill support cartilage, blood vessels, skin, fins, and wound scar tissue (see Figure 2.38).

Vitamin C has many other metabolic roles. Bone calcification is impaired when vitamin C is deficient. It is necessary in iron metabolism, probably to convert transferrin iron from oxidized to reduced form for metabolic transport. It is required in conversion of folic to folinic acid. It is required in tyrosine metabolism; a deficiency causes tyrosine excretion in the urine. Vitamin C deficiency increases blood clotting time. It can spare vitamin E in reducing peroxidation of lipid cellular and subcellular membranes. It is essential for maximum rate of immune responses, and has a role in detoxification of various xenobiotics.

Curvature of the spinal column is a prominent, early sign of vitamin C deficiency in finfishes. Scoliosis and lordosis (lateral and vertical curvature of spinal column, respectively) have been produced by feeding vitamin C-deficient diets to rainbow trout, brook trout, coho salmon, tilapia, channel catfish, and young carp (see Figure 2.39). Lim and Lovell (1978) described the pathology of vitamin C deficiency syndrome in channel catfish as deformed spinal columns, external and internal hemorrhages, erosion of fins, depigmented vertical bands around the midsection, distorted gill filament cartilage, and reduced rate of wound healing. Deformed head and gill operculums occur in

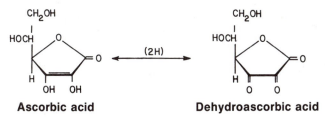

Fig. 2.37. Two forms of vitamin C (ascorbic acid).

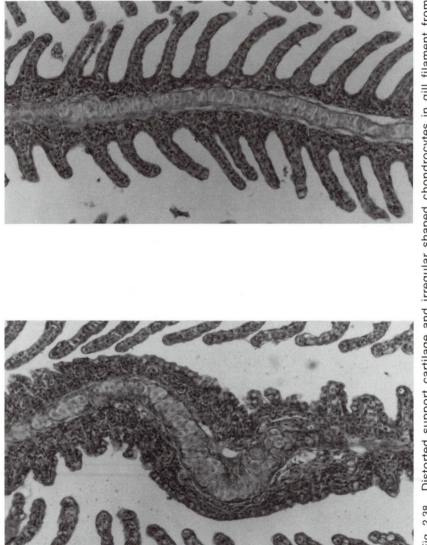

Fig. 2.38. Distorted support cartilage and irregular shaped chondrocytes in gill filament from channel catfish fed diet deficient in vitamin C (left). Gill support from control fish is on the right.

Fig. 2.39. Dietary vitamin C deficiency causes scoliosis and lordosis (lateral and vertical curvature of spinal column) in fish. At bottom is channel catfish fed vitamin C deficient diet for 8 wk; at top is control.

rainbow trout deprived of dietary vitamin C (see Figure 2.40). Lightner et al. (1979) demonstrated that without sufficient vitamin C in the diet, penaeid shrimp died of "black death," a condition characterized by melanized hemocytic lesions distributed throughout the collagenous tissues.

Dietary requirement of vitamin C varies with metabolic function. Hilton, Cho, and Slinger (1978) found that 20 mg of vitamin C per kg of diet was sufficient for normal growth in rainbow trout, but 40 mg/kg was necessary to prevent gross deficiency signs. Li and Lovell (1984) found similar results with channel catfish. Halver, Ashley, and Smith (1969) reported that 50 mg/kg (the lowest level fed) was sufficient for normal growth and bone development in coho salmon, but 400 mg/kg were required for maximum rate of wound healing. Sublethal levels of various pesticides in water increase the requirement of vitamin C by several fishes. Increasing dietary vitamin C reduced incidence of vertebral damage and concentration of toxaphene in the fish, indicating that vitamin C is a factor in detoxification of toxaphene.

Fig. 2.40. Deformed head and gill operculum of vitamin C-deficient rainbow trout.
(Courtesy of John E. Halver)

Dietary requirement for vitamin C by fish seems to decrease with age. Sato, Yoshinaka, and Yamamoto (1978) found that young trout (6 weeks) fed vitamin C-free diets grew slowly and developed scurvy, whereas older trout (19 months) developed none of these problems. Li and Lovell (1984) found that 60 mg/kg of diet was required for normal growth and bone development for small (10 g) channel catfish, but 30 mg/kg was sufficient for larger catfish (50 g).

Rainbow trout can use ascorbate-2-sulfate as a dietary replacement for ascorbic acid; channel catfish can use ascorbate-2-sulfate, but not as efficiently as trout. Tucker and Halver (1984) contend that ascorbate-2-sulfate is a storage metabolite of ascorbic acid in fish which functions to regulate the size of tissue ascorbic acid pools; interconversion of ascorbic acid to ascorbate-2-sulfate is catalyzed by the enzyme ascorbic acid-sulfatase, which is modulated by tissue levels of ascorbic acid through feedback inhibition. Guerin (1986) found, however, only small amounts of ascorbate-2-sulfate in channel catfish, and these concentrations were independent of diet or tissue levels of vitamin C.

Fish fed diets deficient in vitamin C have reduced resistance to bacterial diseases. Elevated dietary doses of ascorbic acid increased resistance of channel catfish to bacterial infections (Li and Lovell 1984). When channel catfish were fed diets containing 0 mg to 3,000 mg of ascorbic acid per kilogram of diet for 13 weeks and subsequently infected with the pathogenic bacterium *Edwardsiella ictaluri,* mortality over 8 days ranged from 100% with the ascorbic acid–deficient diet to none at the highest level of ascorbic acid (Table 2.7). Thirty milligrams per kilogram of the vitamin was sufficient to prevent scurvy in the fish and 150 mg/kg significantly reduced mortality. The immune responses, antibody production, and complement activity were significantly suppressed by omission of ascorbic acid from the diet and significantly enhanced when ascorbic acid was increased from 300 mg/kg to 3,000 mg/kg. Tissue levels of vitamin C plateau at dietary vitamin concentrations of 500 mg/kg to 1,000 mg/kg, so maximum resistance to bacterial infection is probably provided within this diet concentration range. Rainbow trout also showed increased resistance to infection from *Vibrio anguilarum* and increased antibody production when dietary vitamin C was increased to 1,000 mg/kg, which is 10 times the normal requirement. The reason that pharmacologic doses of vitamin C enhance immune responses in fish is not presently known.

Commercial feed ingredients are almost completely devoid of vitamin C, so the vitamin must be supplemented into practical feeds. Vitamin C as l-ascorbic acid is highly sensitive to oxidative destruction during processing and subsequent storage of feeds. Approximately 25% of l-ascorbic acid is destroyed during steam pelleting and 50% is lost in extrusion processing; thus, overfortification is necessary unless the

Table 2.7. PERCENTAGE MORTALITY OF CHANNEL CATFISH FED VARIOUS LEVELS OF VITAMIN C AND INFECTED WITH *EDWARDSIELLA ICTALURI*

Levels of vitamin C, mg/kg diet	Percentage mortality within 8 days post-infection
0	100
30	70
60	70
150	35
300	15
3,000	0

Source: Li and Lovell (1984).

vitamin is added after processing. The half-life of l-ascorbic acid in fish feed during post-processing storage under warm weather conditions is approximately 2.5 months. Phosphate and sulfate conjugates of ascorbic acid are much more stable against oxidation during processing and storage than l-ascorbic acid. These compounds have vitamin C activity for fish and will likely replace the less stable form.

Sources of Vitamins

Vitamins most likely to be deficient in commercial fish feeds that contain oilseed meals, animal byproducts, and grains are vitamins C, A, D, niacin, pantothenic acid, riboflavin, and possibly vitamins E and K. Inositol, biotin, folic acid, pyridoxine, and thiamin are widely found in plant feedstuffs and vitamin B_{12} is present in animal byproducts. Vitamin E and choline are usually extracted with the oil or germ from plant seeds, and may be deficient in many commercial feed formulations.

It is possible to formulate a fish feed from commercial ingredients that is adequate in all essential vitamins except vitamin C. Brewer's yeast or distillery dried grains are good sources, and grain byproducts that include the seed coat are fair sources of the water-soluble vitamins (except C); alfalfa meal is a good source of vitamins E and K; fish meal is a good source of B_{12}; and fish liver oil can supply A and D.

Intestinal Synthesis of Vitamins

Microbial synthesis of vitamins in the gut has not been researched well in fish, although significant amounts of B_{12}, folic acid, inositol, biotin, and K are attributed to this source in warmblooded animals. Studies conducted at Auburn University in which the ratios of vitamins to indigestible dry matter (IDM) were measured in the diet and in the feces (removed from rectum) showed significant increases in inositol and vitamin B_{12} in the digestive tract of channel catfish and tilapia. There was no increase in biotin. Addition of antibiotics significantly reduced the ratio of vitamin to IDM in feces. Increase in vitamin B_{12} in the digestive tract of Nile tilapia was about five times higher than in the digestive tract of channel catfish. Significant absorption of intestinally synthesized B_{12} by channel catfish was revealed by feeding the fish ^{60}Co in the diet and recovering radio-labeled vitamin B_{12} in the blood, liver, kidneys, and spleen (Limsuwan and Lovell 1981).

ESSENTIAL LIPIDS

In addition to being a concentrated energy source, lipids have other nutritional functions. They provide a vehicle for absorption of other fat-soluble nutrients such as sterols and vitamins. They play a role in the structure of cell and subcellular membranes. They are components of hormones and precursors for prostaglandin synthesis. Some lipid components such as sterols and certain fatty acids must be provided preformed in the diet and, thus, are essential nutrients for fishes.

Essential Fatty Acids

Homiothermic animals have a dietary requirement for fatty acids with a double (unsaturated) bond in the n-6 position which is at the sixth carbon from the terminal carbon in the fatty acid chain. Several fatty acids with similar n-6 end structures (18:2 n-6, 20:2 n-6, or 20:4 n-6) can satisfy this requirement. Salmonids, however, require n-3 fatty acids. A possible explanation for the difference in fatty acid requirement is that the omega-3 structure permits a greater degree of unsaturation, which is necessary in the membrane phospholipids to maintain flexibility and permeability characteristics in the fish at low temperatures. The explanation is probably valid because it has been shown that some warmwater fish require n-6 or a mixture of n-3 and n-6 fatty acids while marine fishes, who spend much of their lives at low water temperatures, require n-3 fatty acids. Fatty acid requirements of fish may be temperature dependent; several studies have shown that adipose as well as membrane lipids in fish are affected by temperature. There is evidence also that salinity of the water may affect fatty acid requirements.

The essential fatty acid requirement of rainbow trout, which have the ability to elongate and desaturate n-3 fatty acids, is satisfied by either 18:3 n-3 fatty acids or longer n-3 fatty acids, whereas marine fishes cannot chain elongate fatty acids and thus need dietary 20:5 n-3 or 22:6 n-3 fatty acids preformed from the diet to satisfy their fatty acid requirement.

Some warmwater fishes, such as channel catfish, do not appear to be as sensitive to fatty acid deficiency as rainbow trout and other coldwater species. Early studies (Stickney and Andrews 1972) found no specific fatty acid requirement for channel catfish, but did find that high diet concentrations of 18:2 omega-6 fatty acid reduced growth. Later work, however, has shown that menhaden oil (high in n-3

PUFA) produced higher growth rate in channel catfish than beef fat (no n-3 fatty acids). *Tilapia zilli* appear to require n-6 fatty acids, but not n-3. Some shrimp, such as *Palaemon serratus* and *Penaeus indicus,* seem to require both n-3 and n-6 fatty acids. Determining requirements for fatty acids is difficult for fish because the metabolic requirement is very small and fatty acids stored in the body or even carried over from the egg yolk can influence performance of the experimental fish. Fish used in fatty acid experiments should be carefully depleted of the fatty acids to be tested prior to the experiment.

Fatty acid deficiency signs in rainbow trout were reduced growth, elevated muscle water, increased susceptibility to bacterial infection, increased permeability of mitochondrial membranes, fatty degeneration of livers, and a decrease in hemoglobin in red blood cells. Deficient fish exhibited a fainting or shock syndrome.

Most fish respond to lipid in the diet. How much of this response is caused by supplying essential fatty acids and how much is due to a readily available energy source is difficult to discern. The dietary requirement for n-3 fatty acid, for species that require this form, seems to be about 0.5% to 1% if the fatty acids are greater than 18-carbon (20:5 and 22:6) or up to 2.5% if the n-3 fatty acid is 18-carbon. For fishes requiring n-6 fatty acids or a combination of n-3 and n-6, the requirement appears to be in the range of 0.5% to 1%. For a comprehensive discussion of essential fatty acids in fishes, the reader should refer to Watanabe (1982).

Fatty acid composition of adipose lipid in fish is influenced primarily by diet, whereas membrane lipids are more characteristic of environment and species. Land plants synthesize primarily fatty acids that are 18-carbon or less in chain length with the unsaturated bonds in n-6 and n-9 positions, so "grain-fed" fish will store primarily these types. Some plant oils, however, such as linseed or soybean, contain significant amounts of n-3 fatty acids. Sources of longer chain, n-3 fatty acids (C20:5 C22:6) are primarily freshwater or marine algae, and fish obtain these fatty acids through the food chain. Thus, oil of fish consuming marine or freshwater algae is the primary source of n-3 highly unsaturated fatty acids for fish or human diets.

Sterols and Phospholipids

Finfish readily synthesize sterols from acetate and mevalonic acid; however, crustaceans have limited ability to do this and therefore have a dietary requirement for preformed sterols. The absence of sterol from crustacean diets results in mortality in a short time. The dietary requirement for sterol (as cholesterol) is around 0.5% for penaeid shrimp and lobsters. Marine crustaceans seem to also require the phospholipid lecithin in their diet for maximum growth. Growth rate of penaeid shrimp was improved by adding 1% lecithin to the diet, and growth and survival of lobsters were improved with a 7% supplement of soybean lecithin.

MINERALS

Not all inorganic elements found in an animal's body are essential in its diet. However, dietary need for 22 minerals has been demonstrated in one or more animal species. Those required in large quantities are termed *major* and those required in trace quantities are called *trace* minerals. The major minerals are calcium, phosphorus, magnesium, sodium, potassium, chlorine, and sulfur. Trace minerals are iron, iodine, manganese, copper, cobalt, zinc, selenium, molybdenum, fluorine, aluminum, nickel, vanadium, silicon, tin, and chromium.

The mineral requirements of fish have been studied only sparsely. In addition to the general problems encountered in mineral nutrition research, such as formulating mineral-free diets and overcoming tissue stores of minerals, fish absorb dissolved minerals from the water.

A major difference between mineral metabolism in fish and land animals is osmoregulation, or maintenance of osmotic balance between body fluids in the fish and the water around the fish. Other biochemical functions of minerals in fish are similar to those in warmblooded animals. Some minerals are constituents of hard tissues such as bone, fins, and scales, and some are components of soft tissues, such as sulfur in protein and iron in hemoglobin. Some minerals function as components or activators of enzymes and hormones, such as zinc, which activates alkaline phosphatase and iodine, which is a component of the hormone thyroxine. Some soluble elements, such as calcium, sodium, potassium, and chloride have functions in the blood or body fluids such as osmoregulation, acid-base balance, and sensitizing muscle fibers.

Requirements

Fish probably require the same minerals as warmblooded animals for tissue formation and various metabolic processes. Table 2.8 shows

Table 2.8.   DIETARY REQUIREMENTS FOR
PHOSPHORUS, MAGNESIUM, ZINC,
MANGANESE, AND SELENIUM IN
CHANNEL CATFISH

Element	Amount
Phosphorus, %	0.45
Magnesium, %	0.04
Zinc, mg/kg[1]	20 or 150
Selenium, mg/kg	0.25
Manganese, mg/kg	2.4
Copper, mg/kg	5
Iron, mg/kg	30

[1] 20 mg/kg is the basal requirement, 150 mg/kg
is recommended in practical fish feeds to compensate for antagonistic effect of divalent
cations and phytic acid in the diet on zinc.
Source: Phosphorus, Lovell (1978); magnesium,
Gatlin, Robinson, and Poe (1982); zinc, selenium,
manganese, copper and iron, Gatlin and Wilson
(1983, 1984a, 1984b, 1986a, 1986b).

mineral requirements for channel catfish. Determinations made with
other species generally agree with these values.

Fish can absorb dissolved minerals from the water across the gill
membrane or, in the case of marine fishes that drink water, through
the digestive tract. Most of the calcium requirement for fishes comes
from the water. In seawater, significant amounts of iron, magnesium,
cobalt, potassium, sodium, and zinc can be obtained from the water.
Berg (1968) provided data that indicated that goldfish obtained 50% to
80% of their calcium from the water when fed a calcium-adequate diet.
Fish require a dietary source of phosphorus to meet their relatively
high metabolic requirement because levels of dissolved phosphorus in
natural waters are relatively low.

Calcium and Phosphorus.   Most of the calcium found in the fish
body, perhaps 99%, is in skeletal tissue and scales. From 20% to 40% of
the total calcium is in scales. During fasting, calcium is resorbed from
the hard tissues for physiological functions. The percentage of calcium
in the whole, fresh (wet) body of finfish ranges from 0.5% to 1% with a
ratio of calcium to phosphorus of 0.7 to 1.6. In addition to its structural
functions in bones and scales, calcium is essential for blood clotting,
muscle function, nerve impulse transmission, osmoregulation, and as a
cofactor during various enzymatic processes.

Approximately 85% to 90% of the phosphorus in fish is in bone and
scales. In bone, phosphorus is complexed with calcium to form apatite,

or tricalcium phosphate. Phosphorus also functions in a variety of organic phosphates, such as constituents of adenosine triphosphate (ATP), deoxyribonucleic acid (DNA), ribonucleic acid (RNA), various coenzymes, and phospholipids in cell and subcellular membranes. The phosphate buffer system maintains normal pH in body fluids.

Calcium deficiency was produced in channel catfish grown in calcium-free water. This was characterized by reductions in growth and ash content of bones. Signs of dietary phosphorus deficiency for most fishes include poor growth and bone mineralization. Further signs of deficiency observed in carp include increases in carcass fat, reduced blood phosphate levels, deformed head (frontal bone), abnormal calcification of ribs and soft rays of the pectoral fins, and a curved or abnormal spine. Symptoms of phosphorus deficiency in red sea bream include deformed vertebrae and increased serum alkaline phosphatase, with increased fat content and decreased glycogen content in the liver.

Minimum requirement of available phorphorus in diets foₗ channel catfish was determined to be 0.45% with purified diets. Requirements for common carp, Nile tilapia, red sea bream, and eel have been determined to be 0.6%, 0.9%, 0.68% and 0.29% of the diet, respectively.

Phytate phosphorus (approximately 67% of the phosphorus in grains is in the phytate form) is poorly available to fish. Phosphorus in fish meal is 40% to 75% available to fish with gastric stomachs, but less than 25% available to the stomachless carps. Inorganic phosphorus from sodium or monocalcium phosphate is highly available to all fish; however, dicalcium phosphate is less available. Availability of phosphorus in several feed ingredients for catfish, carp, and trout is presented in Table A.3 in Appendix A (see page 248).

Magnesium.   About 70% of the magnesium in a fish's body is in the hard tissue. Other functions of magnesium are as an enzyme activator in carbohydrate metabolism and in protein synthesis. Magnesium is necessary in body fluids to maintain integrity of smooth muscle. Deficiency causes tetany in warmblooded animals and flaccid muscle in fish.

Magnesium deficiency in the diet of channel catfish causes poor growth, anorexia, lethargy, flaccid muscles, high mortality, and depressed magnesium levels in the whole body, blood serum, and bones. Deficiency of magnesium in the diet of common carp caused similar signs, in addition to convulsions. Signs of deficiency in rainbow trout include mortality, vertebral curvature, degeneration of muscle fibers, and degeneration of the epithelial cells of the pyloric caecae and gill filaments. Red sea bream reared in seawater, which typically contains high levels of magnesium, showed no signs of deficiency. Fish in fresh water, which only contains 1 mg/L to 3 mg/L of magnesium, require 0.025% to 0.07% magnesium in the diet.

Most foods, especially plants, are high in magnesium, which usually makes it unnecessary to supplement feeds made from natural ingredients; however, purified diets must have a magnesium supplement.

Iron. The principal role of iron in a fish's body is as a component of hemoglobin. Another role is as a component of the cytochrome enzyme system that regenerates ATP in cellular oxidation.

Hemoglobin is the oxygen-carrying pigment in red blood cells. Red blood cells are formed primarily in the spleen and anterior kidney in fish instead of the bone marrow, as in land animals. Red blood cells are regenerated periodically and most of the iron is recycled. That which is not recycled is excreted through the bile into the intestine. A deficiency of iron can cause anemia.

Iron, like other elements of low solubility, such as zinc and copper, is absorbed and transported in the body in protein-bound form. In the intestinal cell mucosa, iron combines with a protein, apoferritin, to form ferritin. The amount of apoferritin in the mucosa is regulated by body need for iron. Iron is in the oxidized form (Fe^{+++}), when combined with the protein in the mucosa. When liberated into the blood, it is reduced to Fe^{++} by reducing agents, such as vitamin C. It is transported in the blood bound to another protein as transferrin and stored in the liver and hemopoietic tissues as Fe^{+++} in combination with a protein until used. Iron and other minerals of low solubility are not excreted through the urine, but are returned to the digestive tract.

Dietary deficiency of iron causes microcytic anemia in fish. Dietary requirement for iron by channel catfish is 30 mg/kg of diet (Gatlin and Wilson 1986b). Dissolved iron in the water can serve as a source of iron for fish metabolism, although dissolved iron often precipitates out as ferric hydroxide and levels in solution are low. Iron is widespread in feedstuffs; however, its availability from plant feeds is relatively low. Unless fish feeds contain significant amounts of animal byproducts, supplemental iron should be used.

Copper. Copper is involved with iron absorption and metabolism. When the diet is deficient in copper, iron levels in body tissues decrease. Copper functions in hematopoiesis (hemoglobin formation) and in several enzyme systems, such as cytochrome C oxidase and tyrosinase. It is essential in bone development, perhaps through its role in collagen synthesis. Some sea animals, molluscs, and crustaceans contain copper as the metal nucleus of the oxygen-carrying pigment in the blood, hemocyanin or cyanodin, which has an analogous role to hemoglobin in red-blooded animals. Like iron, copper is absorbed and transported as a copper-protein complex.

Common carp require approximately 3 mg of copper per kilogram of diet for normal growth. Channel catfish require 1.5 mg/kg to 5 mg/kg for normal growth and blood cell formation, but 32 mg/kg caused growth depression and anemia. This indicates a rather narrow range of dietary copper tolerance. Approximately 100 mg to 250 mg copper/kg of diet is toxic to livestock.

Iodine.   Iodine is a component of thyroxin, the thyroid hormone that regulates rate of metabolism. If the amino acid tyrosine and iodine are supplied, the body can synthesize thyroxin. Deficiency of iodine results in hyperplasia of the thyroid gland, or goiter, in fishes. Reports have shown that iodine can be absorbed directly from the water by fish. Iodine is absorbed efficiently from the gut and is transported in the body bound to a protein. It is excreted in the urine.

The minimum requirement of fish for dietary iodine has not been defined; however, 1 mg/kg to 5 mg/kg feed has been found to be an adequate level. Fish meal is a rich source of iodine. Fish feeds not containing fish meal should probably be supplemented with iodine in the absence of natural aquatic foods.

Zinc.   Zinc has several functions. It serves as a cofactor in several enzyme systems, including carbonic anhydrase found in red blood cells, enzymes in protein digestion, and enzymes in carbohydrate catabolism. It plays a role in preventing keratinization of epithelial tissues. Insulin is stored as a zinc complex.

Gatlin and Wilson (1983) demonstrated that zinc-deficient channel catfish had depressed growth, appetite, and serum alkaline phosphatase activity, and reduced levels of zinc and calcium in bones. They showed that channel catfish had a minimum dietary zinc requirement of 20 mg/kg to prevent deficiency signs, but the requirement for optimum growth was lower. Zinc deficiency in common carp caused slow growth, loss of appetite, high mortality, and erosion of the skin and fins. Dietary levels of 5 mg/kg allowed maximum growth rate, but 15 mg/kg to 30 mg/kg were required to prevent all deficiency signs.

Zinc deficiency produces cataracts in rainbow trout. Ketola (1979) produced bilateral lens cataracts in rainbow trout by feeding diets with higher than normal requirement of zinc, but which contained fish meal with high bone ash. He corrected the problem by supplementing 150 mg/kg zinc into the diet, and speculated that the high level of calcium or other minerals in the fish meal impaired zinc absorption. Over 100 mg/kg dietary zinc is recommended in fish feeds containing soybean meal to compensate for the zinc-binding property of phytic acid.

Manganese. Manganese functions as a cofactor in several enzyme systems, including those involved in synthesis of urea from ammonia, amino acid metabolism, fatty acid metabolism, and glucose oxidation.

Manganese-deficient diets caused depressed growth in common carp and rainbow trout, and abnormal tail growth and shortening of the body in the latter. Supplementation of the diet to bring the level of manganese to 12 mg/kg to 13 mg/kg improved growth in both species and prevented abnormalities in rainbow trout. However, Gatlin and Wilson (1984b) found that 2.4 mg manganese/kg diet was sufficient for normal growth and health of channel catfish.

Selenium. The most notable function of selenium is as a component of the enzyme glutathione oxidase, which acts along with vitamin E as a biological antioxidant to protect polyunsaturated phospholipids in cellular and subcellular membranes from oxidation damage. Selenium has also been identified as a cofactor in glucose metabolism.

Selenium deficiency in Atlantic salmon caused increased mortality and suppressed plasma glutathione peroxidase activity. Either selenium or vitamin E deficiency caused nutritional muscular dystrophy. Maximum plasma glutathione peroxidase activity occurred in rainbow trout fed diets containing 0.38 mg/kg of selenium. A level of 13 mg/L selenium in trout diets caused suppressed growth and increased mortality. Gatlin and Wilson (1984a) found that maximum growth and glutathione peroxidase activity occurred in channel catfish fed diets containing 0.25 mg of selenium/kg of diet, and that 15 mg of selenium/kg was toxic to channel catfish.

In the United States, selenium is routinely supplemented in corn-soybean meal feeds for poultry and swine to compensate for deficiencies in feedstuffs produced on selenium-deficient soils. Fish feeds containing predominately plant ingredients should contain a selenium supplement.

Sodium, potassium, and chloride. Dietary deficiencies of sodium, potassium, and chloride have not been produced in fishes, although these elements are necessary for osmoregulations and pH balance in the body fluids, nerve impulse transmissions, and other functions.

Sodium and potassium are the major extracellular and intracellular cations, respectively, and chloride is the major extracellular anion. The chloride ion is a component of hydrochloric acid, which is secreted in the stomach. Most freshwater and all seawater probably contains sufficient amounts of these three ions to satisfy the physiological needs of fish. Fish absorb the ions through the gills in freshwater and through the gut in seawater. Channel catfish did not respond to the addition of salt (NaCl) to salt-free purified diets fed in freshwater. Because most fish can excrete large dietary intakes of salt effectively, dietary levels of 8 to 12 percent salt have had no adverse effect on subadult or adult fish fed in fresh or seawater.

REFERENCES

ARAI, S. 1986. Personal communication. National Research Institute of Aquaculture, Tamaki, Mie, Japan.

BERG, A. 1968. Studies on the metabolism of calcium and strontium in freshwater fish. I. Relative contribution of direct and intestinal absorption. *Mem. Ist Ital. Idrobiol. Dott. Marco de Marchi* 23: 161–196.

BRAMBILA, S. and F. W. HILL. 1966. Carbohydrate requirement of chickens. *J. Nutr.* 88: 84–89.

BRODY, S. [1945] 1974. *Bioenergetics and growth.* New York: Reinhold. Reprinted, New York: Hafner Press.

BURTLE, G. J. 1981. Essentiality of dietary inositol for channel catfish. Ph.D. diss., Auburn University, Auburn, AL.

BUTTERWORTH, C. E., JR., J. A. PLUMB, and J. M. GRIZZLE. 1986. Abnormal folate metabolism in feed-related anemia in cultured catfish. *Proc. Soc. Exp. Biol. Med.* 181: 210–216.

CASTLEDINE, A. J., C. Y. CHO, S. J. SLINGER, B. HICKS, and H. S. BAYLEY. 1978. Influence of dietary biotin on growth, metabolism and pathology of rainbow trout. *J. Nutr.* 108: 698–711.

CHO, C. Y., S. T. SLINGER, and H. S. BAYLEY. 1982. Bioenergetics of salmonid fishes: Energy intake, expenditure and productivity. *Comp. Biochem. Phys.* 73B: 25–41.

COWEY, C. B., E. DEGENER, A. G. T. TACON, A. YOUNGSON, and J. G. BELL. 1984. Effect of vitamin E and oxidized fish oil on the nutrition of rainbow trout grown at natural and varying water temperatures. *Brit. J. Nutr.* 51: 443–451.

GARLING, D. L., JR., and R. P. WILSON. 1977. Effects of dietary carbohydrate to lipid ratios on growth and body composition of fingerling channel catfish. *Prog. Fish-Cult.* 39: 43–47.

GATLIN, D. M., III, E. H. ROBINSON, and W. E. POE. 1982. Magnesium requirement of fingerling channel catfish. *J. Nutr.* 112: 1181–1187.

GATLIN, D. M., III, and R. P. WILSON. 1983. Dietary zinc requirements of channel catfish. *J. Nutr.* 113: 630–635.

GATLIN, D. M., III, and R. P. WILSON.   1984a.   Dietary selenium requirement of fingerling channel catfish. *J. Nutr.* 114: 627–633.

GATLIN, D. M., III, and R. P. WILSON.   1984b.   Studies on manganese requirement of fingerling channel catfish. *Aquaculture* 41: 85–92.

GATLIN, D. M., III, and R. P. WILSON.   1986a.   Dietary copper requirements for channel catfish. *Aquaculture* 54: 277–285.

GATLIN, D. M., III and R. P. WILSON.   1986b.   Characterization of iron deficiency and dietary requirement for channel catfish. *Aquaculture* 52: 191–198.

GUERIN, M.   1986.   Effects of feeding channel catfish various levels of vitamin C in ponds on resistance to *Edwardsiella ictaluri* infection and tissue levels of vitamin C. Master's thesis, Auburn University, Auburn, AL.

HALVER, J. E., L. M. ASHLEY, and R. R. SMITH.   1969.   Ascorbic acid requirements of coho salmon and rainbow trout. *Trans. Am. Fish. Soc.* 98: 762–771.

HILTON, J. W., C. Y. CHO, and S. J. SLINGER.   1978.   Effect of graded levels of supplemental ascorbic acid in practical diets of rainbow trout. *J. Fish. Res. Bd. Can.* 35: 431–436.

KASHIWADA, K., S. TESHIMA, and A. KANAZAWA.   1970.   Studies on the production of B vitamins by intestinal bacteria of fish. V. Evidence of the production of vitamin B_{12} by microorganisms in the intestinal canal of carp, *Cyprinus carpio. Bull. Jpn. Soc. Sci. Fish.* 36: 421–424.

KETOLA, H. G.   1979.   Influence of dietary zinc on cataracts in rainbow trout (*Salmo gairdneri*). *J. Nutr.* 109: 965–969.

LEE, PING-PING H.   1987.   Metabolism of carotenoids by channel catfish. Ph.D. diss., Auburn University, Auburn, AL.

LI, Y., and R. T. LOVELL.   1984.   Elevated levels of dietary ascorbic acid increase immune responses in channel catfish. *J. Nutr.* 115: 123–131.

LIGHTNER, D. V., B. H. Huner, P. C. MAGERELLI, and L. B. CALVIN.   1979.   Ascorbic acid: Nutritional requirement and role in wound repair in shrimp. *Proc. World Maric. Soc.* 9: 447–458.

LIM, C., and R. T. LOVELL.   1978.   Pathology of the vitamin C deficiency syndrome in channel catfish (*Ictalurus punctatus*). *J. Nutr.* 108: 1137–1146.

LIMSUWAN, T., and R. T. LOVELL.   1981.   Intestinal synthesis and absorption of vitamin B_{12} in channel catfish. *J. Nutr.* 111: 2125–2132.

LOVELL, R. T.   1978.   Dietary phosphorus requirement of channel catfish. *Trans. Am. Fish. Soc.* 107: 617–621.

LOVELL, R. T.   1984.   Use of soybean products in diets for aquaculture species. *Res. Highlights, Amer. Soybean Assoc.*, February.

LOVELL, R. T., T. MIYAZAKI, and S. REBEGNATOR.   1984.   Requirement of alpha-tocopherol by channel catfish fed diets low in polyunsaturated fatty acids. *J. Nutr.* 114: 894–901.

LOVELL, R. T., and R. R. STICKNEY.   1977.   Nutrition and feeding of channel catfish. *So. Coop. Bull.* 218.

MANGALIK, A.   1986.   Dietary energy requirements for channel catfish. Ph.D. diss., Auburn University, Auburn, AL.

MURAI, T.   1985.   Biological assessment of nutrient requirements and availability of fish. Special workshop at the International Congress on Nutrition, August 19–25, Brighton, England.

NATIONAL RESEARCH COUNCIL. 1979. *Nutrient requirements of swine.* Washington, DC: National Academy of Sciences.

NATIONAL RESEARCH COUNCIL. 1981. Nutrient requirements of coldwater fish. Washington, DC: National Academy of Sciences.

NATIONAL RESEARCH COUNCIL. 1983. Nutrient requirements of warmwater fish. Washington, DC: National Academy of Sciences.

NATIONAL RESEARCH COUNCIL. 1984. Nutrient requirements of poultry. Washington, DC: National Academy of Sciences.

PAGE, J. W., and J. W. ANDREWS. 1973. Interactions of dietary levels of protein and energy on channel catfish (*Ictalurus punctatus*). *J. Nutr.* 103: 1339–1346.

POSTON, H. A., R. C. RIIS, G. L. RUMSEY, and H. G. KETOLA. 1977. The effect of supplemental dietary amino acids, minerals and vitamins on salmonids fed cataractogenic diets. *Cornell Vet.* 67: 472–509.

PRATHER, E. E., and R. T. LOVELL. 1973. Response of intensively-fed catfish to diets containing various protein-to-energy ratios. *Proc. 27th Ann. Conf. Southeast. Assoc. Game Fish Comm.* 27: 455–459.

ROBBINS, K. R., H. W. NORTON, and D. H. BAKER. 1979. Estimation of nutrient requirements from growth data. *J. Nutr.* 109: 1710–1714.

SANTIAGO, C. B. 1985. Amino acid requirements of Nile tilapia. Ph.D. diss., Auburn University, Auburn, AL.

SATO, M., R. YOSHINAKA, and S. YAMAMOTO. 1978. Nonessentiality of ascorbic acid in the diet of carp. *Bull. Jpn. Soc. Sci. Fish.* 44: 1151–1156.

SMITH, R. R., G. L. RUMSEY, and M. L. SCOTT. 1978. New energy maintenance requirements of salmonids as measured by direct calorimetry: Effect of body size and temperature. *J. Nutr.* 108: 1017–1024.

STICKNEY, R. R., and J. W. ANDREWS. 1972. Effect of dietary lipids on growth, food conversion, fatty acid composition of channel catfish. *J. Nutr.* 102: 249–258.

TAKEUCHI, T., T. WATANABE, and C. OGINO. 1979. Optimum ratio of energy to protein for carp. *Bull. Jpn. Soc. Sci. Fish.* 45: 983–987.

TUCKER, B. W., and J. E. HALVER. 1984. Ascorbate-2-sulfate metabolism in fish. *Nutr. Rev.* 45: 173–179.

WATANABE, T. 1982. Lipid nutrition in fish. *Comp. Biochem. Physiol.* 73B: 3–15.

WILSON, R. P., E. H. ROBINSON, and W. E. POE. 1981. Apparent and true availability of amino acids from common feed ingredients for channel catfish. *J. Nutr.* 111: 923–929.

WILSON, R. P., P. R. BOWSER, and W. E. POE. 1984. Dietary vitamin E requirement for fingerling channel catfish. *J. Nutr.* 114: 2053–2058.

ZIETOUN, I. H., D. E. ULBREY, and W. T. MAGEE. 1976. Quantifying nutrient requirements of fish. *J. Fish. Res. Bd. Can.* 33: 167–172.

3

Digestion and Metabolism

DIGESTION

Digestion may be defined as the preparation of food by the animal for absorption. As such, this may include mechanical reduction of particle size (grinding by pharyngeal teeth or gizzard), enzyme solubilization of organics, pH solubilization of inorganics, and emulsification of lipids. Absorption includes the various processes that allow ions and molecules to pass through membranes of the intestinal tract into the blood or lymph to be metabolized by the animal.

The major divisions of the vertebrate digestive tract are mouth, esophagus, pharynx, stomach, intestine, rectum, and secretory glands, which include the liver and pancreas. Not all fish have functionally divided parts—in some fish, the digestive tract is a long tube from mouth to anus. Usually, however, the esophagus,, stomach, and intestine can be distinguished. The principal layers of the wall of the digestive tract are the mucosa (or inner epithelium), submucosa (mainly connective tissue and blood vessels), the muscularis (two or three layers of smooth muscle), and the serosa or outside layer (fibrous connective tissue).

Fish vary tremendously in morphology and physiology of digestive tracts, and in feeding behavior. Most fishes do not have teeth or gizzards for grinding food; examples of exceptions are the grass carp (which has pharyngeal teeth) and the gizzard shad. Crustaceans grind food prior to ingestion. The most pronounced distinction among fishes with different digestive processes is that some have a gastric section (stomach) and some have none. Some fish have digestive tracts less than one-half the length of their body and others have tracts six to eight times their body lengths. Some feed exclusively on plankton concentrated from the water, others are strictly bottom feeders, and many species feed only on large prey animals. It is indeed hazardous to make a generalization about the digestive system and feeding behavior of fishes. Figure 3.1 shows the digestive tracts of carp, which has no stomach, and channel catfish, which has a well developed gastric section.

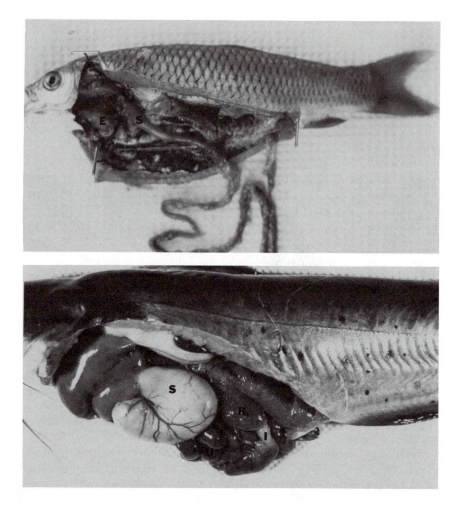

Fig. 3.1. The grass carp (top) has no functional stomach and a relatively long digestive tract: (E) esophagus; (S) poorly defined stomach area. The channel catfish (bottom) has a highly functional stomach and relatively short digestive tract: (S) stomach; (U) upper intestine, (I) intestinal sphincter; (R) rectal intestine.

Mouth and Esophagus

The channel catfish has a large mouth and esophagus for capturing prey (catfish are only slightly predaceous when supplemental feed is offered); the mouth has no teeth but an abrasive plate and there are pads but no teeth in the pharynx; there is no gizzard; the esophagus is separated from the stomach by a cardiac sphincter which effectively separates water from ingested food and prevents food from backing out of the stomach; gill rakers are not arranged for filter feeding; and, catfish have chemosensory barbels that aid in finding food.

The common carp has a small mouth designed for bottom feeding, moderately developed pharyngeal teeth, two pairs of chemosensory barbels, and coarse gill rakers. The grass carp has large pharyngeal teeth attached to branchial bones which serve as grinders to reduce plant tissues to smaller particles sizes.

Tilapias have mouths intermediate between bottom feeders and predaceous fish. Most tilapias are efficient plankton feeders, although not all have closely spaced gill rakers. An important mechanism for concentrating plankton is secretion of pharyngeal mucus that coalesces the plankton so the fish can swallow it.

Stomach

Channel catfish, salmonids, and most of the fin fishes have true stomachs that secrete hydrochloric acid and pepsinogen. Common carp has no stomach, but a slightly enlarged "bulb" exists at the anterior end of the digestive tract. Bile and pancreatic secretions empty into the carp intestine just posterior to the cardiac sphincter which separates the esophagus from the intestine. There is no gastric (low pH) section in the carp's gut.

Tilapias have a modified stomach which secretes hydrochloric acid. There is a well defined compartment, however, which appears to be more of a pocket through which ingesta may divert as it moves through the gut rather than a specialized section, as in catfish. The pH of the ingesta in the stomach is dependent upon the amount of food passing through the gut. When the stomach is empty, the first food eaten may reach pH 1.4 but as more is eaten, gastric secretion is unable to maintain the low pH. In stomached fish and higher animals, the pyloric sphincter (separates stomach from intestine) valve generally remains closed until the stomach contents (chyme) reach a sufficiently low pH to release ingesta into the lower intestine.

Intestine

The length of the digestive tract in the channel catfish is less than the length of the body of the fish. The intestine is not separated into a large and small intestine as in mammals. The catfish's intestine, except for a

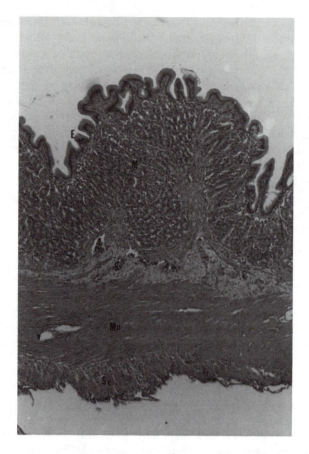

Fig. 3.2. Histological section of walls of stomach (left) and intestine (right) of channel catfish. The sections, from inside outward, are epithelium (E), mucosa (M), submucosa (S), muscularis (Mu) and serosa (Se). Note the thick mucosa, due to presence of gastric glands and thick muscularis, in the stomach, and the large amounts of folding and epithelial surface in the intestine (Grizzle and Rogers 1976).

short rectum, is similar with regard to pH (slightly above 7), digestive secretions, and nutrient absorption to the small intestine in warm-blooded animals. Although the intestine is relatively short, it has many folds, which provide a large surface area for absorption (Figure 3.2).

The entire digestive tract of common carp, which is about three times the length of its body, is like the intestine (posterior to the stomach) of the catfish in alkalinity, secretions, and absorptive functions. Tilapias have a long digestive tract that is six to eight times the

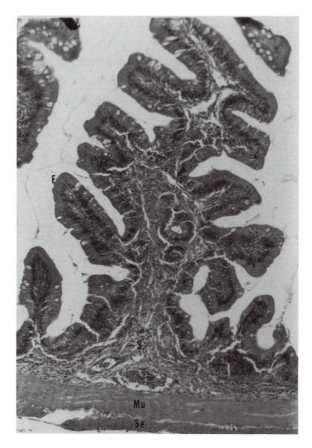

Fig. 3.2. *(continued)*

length of their bodies. The intestine posterior to the stomach is similar to that of the carp except for being longer.

Liver and Pancreas

These organs produce digestive secretions. The liver produces bile in addition to being the primary organ for synthesis, detoxification, and storage for many nutrients. The pancreas is the primary source of digestive enzymes (found as precursors or "zymogens") in most animals. In teleost fishes the pancreatic tissue is usually quite diffuse. It consists of acini (ramified tubules) scattered in the mesenteries, along the intestinal surface, and within the liver and spleen. Crustaceans have a hepatopancreas, a distinct organ that functions as liver and pancreas.

Digestive Processes

Digestive processes in the channel catfish, beyond the mouth, are relatively similar to those in warmblooded monogastric animals; these will be described and exceptions in other cultured fishes will be mentioned.

Stomach. When food enters the stomach, neural and hormonal processes stimulate digestive secretions. Distension of the stomach by entering food starts hydrochloric acid secretion by parietal cells. Pepsinogen is secreted by chief cells and is quickly hydrolyzed to pepsin, an active proteolytic enzyme. Mucous cells begin to secrete. In mammals, the hormone gastrin is secreted into the stomach to stimulate release of gastric juices; however, in fishes, it is believed that other hormones or compounds (histamine, cerulein) may be more functional. The pyloric sphincter at the posterior end of the stomach holds food until it is sufficiently fluid to be passed into the small intestine. Water must be secreted into the stomach of freshwater fishes, but saltwater fishes drink water. Pepsin and hydrochloric acid partially hydrolyze proteins into shorter chain polypeptides. Pepsin, which is active at pH 1.5 to 3.0, works mainly at the site of aromatic amino acids (phenylalanine and tyrosine). Minerals and mineralized tissue are solubilized in the acid stomach, but no fat or carbohydrate breakdown occurs. The mixture of food, mucus, and gastric juices becomes a slurry, called chyme, which when sufficiently fluid is released gradually into the anterior intestine.

Intestine. In mammals, chyme entering the intestine initiates release of secretions from the pancreas and gall bladder (bile). Pancreatic secretions include bicarbonate buffering compounds, which neutralize the acidic chyme: and the zymogens of enzymes, which digest proteins, carbohydrates, lipids, chitin, and nucleotides. Trypsinogen, the zymogen of trypsin, is activated by enterokinase, which is secreted by the intestinal mucosa. Chymotrypsinogen also comes from the pancreas and is activated to chymotrypsin when it comes in contact with trypsin. These two proteolytic enzymes cleave polypeptides into shorter chain peptides. Carboxypeptidases and aminopeptidases may also come from the pancreas. They split off individual amino acids, containing the free carboxyl or the free amino group, from a peptide chain.

Chitinase activity has been found in pancreas extracts from many fishes. This enzyme hydrolyzes the chitinous exoskeleton of insects and crustaceans. The pancreas also produces amylase, which hydrolyzes starch, and nucleases, which degrade nucleic acids. Pancreatic lipases split ester bonds and partially or completely hydrolyze fats, phospholipids, and other lipid esters. Cellulase activity has been reported in intestinal extracts from a few fishes, but this probably came from intestinal bacteria.

Cells of the intestinal mucosa of most vertebrates secrete a number of enzymes. These include carboxy- and aminopeptidases, amylases, lipases, licethinase, nucleases, and others. The quantitative importance of intestinal mucosa enzymes to digestion in fish is not known. Grizzle and Rogers (1976) found no secretory glands in the intestine of channel catfish.

Bile is produced in the liver and is secreted into the intestine, usually via the gall bladder. It contains bile salts, cholesterol, phospholipids, pigments, and other compounds and ions. Its primary function is to emulsify fats into small globules (chylomicrons) for absorption or to make hydrolysis by lipases easier.

Food moves through the digestive tract by peristaltic waves or constrictions that move along the intestine. The food is churned in the foregut by various independent movements of the layers of smooth muscle in the intestinal wall. This aids in exposing the food to the digestive secretions and also in exposing the nutrients to the intestinal mucosa for absorption.

Most nutrient absorption occurs in the intestine. In the lumen of the intestine, there are many folds to provide for a large surface area for nutrient absorption (Figure 3.2). Some highly soluble nutrients, such as electrolytes, monosaccharides, and some of the vitamins and amino acids, diffuse across the mucosal cell membrane because of concentration gradient (passive diffusion). Other nutrients must be actively transported into the cell. Another mechanism of cell absorption is pinocytosis, where the cell engulfs large molecules in amoeba-like fashion.

Proteins are absorbed primarily as free amino acids, but also as low molecular weight peptides. Triglycerides are absorbed as small fat particles (micelles) and as free fatty acids and glycerol.

Water-soluble vitamins are absorbed free, but fat-soluble vitamins are solubilized in a lipid medium when absorbed. Carbohydrates are absorbed as glucose or other monosaccharides. Electrolytes are absorbed free, calcium and phosphorus are complexed together, and most of the trace minerals (iron, copper, zinc, etc.) are coupled with proteins to pass into the mucosal cells.

All nutrients, except possibly lipids, are absorbed from the mucosal cells into the blood that moves from the intestine through the portal vein to the liver. In birds and mammals, lipids are picked up by the lymph as triglycerides, which are formed after absorption into the mucosal cells, and later enter the blood for transport to the liver. Teleost fishes have a well defined lymph system, but the system's role in lipid absorption in fish is not clear.

Fishes do not have a well defined large intestine, as do higher animals, where bile salts, some minerals, most of the water, and nutrients synthesized in the intestine are absorbed. These processes apparently do occur in fishes in the posterior section of the intestine.

MEASURING NUTRIENT DIGESTIBILITY IN FISH

Percentage apparent digestibility (D) of a nutrient is expressed by equation (3.1):

$$\%D = 100 \times \frac{Amount\ of\ nutrient\ fed\ -\ Amount\ of\ nutrient\ in\ feces}{Amount\ of\ nutrient\ fed}$$

(3.1)

Apparent digestibility does not take into account nutrient losses of endogenous origin (secretions from the gut) which are part of the feces. "Corrected" or "true" digestibility calculations exclude the endogenous materials from the feces. The endogenous materials are primarily nitrogenous compounds, such as enzymes, peptides, and epithelial cells, and must be determined in a separate study in which a nitrogen-free diet is fed. Apparent digestibility is of more practical importance than corrected digestibility because the endogenous losses are minor if the animal is not fed, therefore these losses must be charged to ingestion of the food.

Quantitative recovery of the fed nutrient voided in the feces is difficult in fish because of the aqueous environment. Several methods of direct and indirect measurement of digestibility have been used with fish. The direct method involves measuring directly all of the nutrient consumed and all excreted in the feces. Care must be taken when feces are collected from the water so that leaching of nutrients into the water is prevented. Satisfactory methods for making total fecal collections are described by Cho, Slinger, and Bayley (1982).

The indirect method involves measurement of the ratios of nutrient to some indigestible component (indicator) in the feed and in the feces. The indicator must be indigestible, unaltered chemically, nontoxic to the fish, conveniently analyzed, and able to pass through the gut uniformly with other ingesta. As the dietary nutrient is absorbed in the gut, the ratio of nutrient to indicator will be less in the feces than in the feed. Digestibility has been determined indirectly in fishes and shrimps, using internal indicators such as ash, crude fiber or plant chromagens, or diet additives such as chromic oxide. The equation for calculating percentage digestibility by the indirect method is presented in chapter 2 (equation 2.2, page 15). Procedures for measuring digestion indirectly in channel catfish, using chromic oxide as an indicator are described by Smith and Lovell (1971, 1973).

Popma (1982) showed that natural plant chromagens (chlorophyll derivatives) are suitable reference compounds and may be used to measure nutrient digestion by naturally feeding fish when the ingested food contains green plant tissue. He sampled ingested food from the esophagus of Nile tilapia before digestion had commenced and fecal material from the rectal area of the fish from fish free in a pond with access to a variety of foods. It is important that the species or particle composition of the sampled ingesta be identified in order to know what the determined digestion coefficients represent.

METABOLISM

Metabolism may be defined as the biological processes of utilization of absorbed nutrients for growth and other synthesis and for energy expenditure.

Metabolism of Carbohydrates

The protein-sparing effect of carbohydrates varies from poor to relatively good among fish species. Fish have poorer control over blood glucose levels than do warm-blooded animals; in fact, fish respond to glucose loading like a diabetic mammal. Following glucose ingestion, blood glucose level rises rapidly in fish, but takes many hours to decrease. Turnover rate of glucose in trout is about 10 times slower than in the rat. Fish oxidize deaminated amino acids for energy more efficiently than, and in preference to, glucose. Because fish evolved in a carbohydrate-poor environment, it can be appreciated that they do not utilize carbohydrates efficiently, although the reasons are not well understood.

Absorbed carbohydrates, primarily in the form of glucose, may have three major metabolic roles: (a) an immediate source of energy, (b) stored as glycogen as reserve energy, and (c) synthesized into compounds such as triglycerides, nonessential amino acids, and others.

Metabolism of glucose begins with glucose phosphorylation and proceeds with glucose being converted into glycogen or degraded by the Embden-Meyerhof (EM) pathway or the pentose-phosphate pathway (Figure 3.3). The EM pathway takes glucose-6-PO_4 to pyruvate (glycolysis), whereas the pentose-phosphate pathway takes it to ribose or other compounds which go back into the EM pathway and proceed to pyruvate.

Pyruvate penetrates the mitochondrion and is decarboxylated to 2-carbon acetyl, which combines with coenzyme-A (CoA) to form acetyl-CoA in a reaction catalyzed by pyruvate dehydrogenase complex. Acetyl-CoA enters the tricarboxylic acid (TCA) cycle (Figure 3.4) by combining with oxaloacetate. In the TCA cycle, carbon is released as CO_2 and hydrogen is transferred to NAD^+ and FAD. These coenzymes transfer hydrogen atoms to the cytochrome enzyme system (electron transport chain) which in turn promotes the transfer of the hydrogen (electrons) to molecular oxygen, ultimately forming H_2O. As electrons pass through the electron transport chain, energy is released; some is immediately lost as heat and some is captured by oxidative phosphorylation. In this process, energy is retained in the formation of phosphate bonds regenerating adenosine triphosphate (ATP) from ADP and inorganic phosphate. ATP, which provides the energy for endergonic metabolic processes, is composed of a nitrogenous base (adenine), a 5-carbon sugar (ribose), and three phosphate units (Figure 3.5). There are compounds analogous in function to ATP, but less prominent, which contain other nitrogenous bases, such as GTP (guanine), UTP (uracil), and CTP (cytosine).

Six to eight moles of ATP are produced in the glycolysis of 1 mole of glucose to 2 moles of pyruvate. For 2 moles of pyruvate oxidized, 30 moles of ATP are produced. Thus, the net gain from glycolysis and oxidation of 1 mole of glucose is 36 to 38 ATP's. If each mole of ATP represents 7.3 kcal of chemical energy captured through oxidative phosphorylation, the efficiency of energy conversion of glucose, which has a gross energy value of 673 kcal/mole, is about 40%:

$$100 \times \frac{38 \ moles \ ATP \times 7.3}{673} = 41.2\%. \qquad (3.2)$$

Note in Figures 3.3 and 3.4 that keto acids, from deaminated amino acids, and triglycerides are also oxidized through the TCA cycle. Keto acids enter at pyruvate, acetyl-CoA, and several points in the TCA cycle. Fatty acids enter through acetyl-CoA. Glycerol (from hydrolyzed fats) enters through the glycolytic pathway at dihydroxyacetone phosphate.

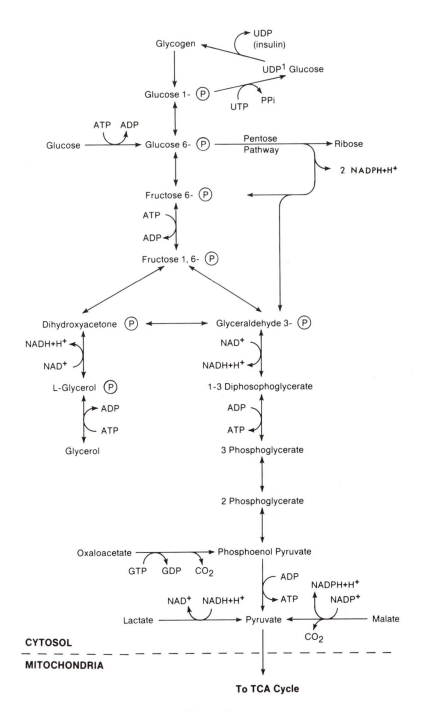

Fig. 3.3. Glycolytic pathway.

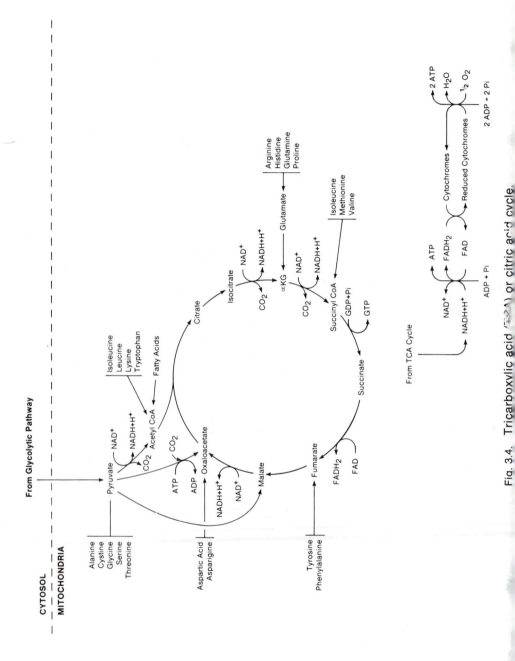

Fig. 3.4. Tricarboxylic acid (TCA) or citric acid cycle.

84

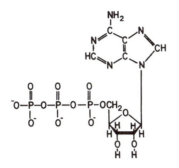

Fig. 3.5.   Adenosine triphosphate (ATP)

All of the enzymes for glycolysis have been found in fish tissues. The reader is referred to biochemistry textbooks for a discussion of these enzymes. Heart and white muscle usually have the highest rate of activity, followed by brain, kidney, gills, and liver. The role of glycolysis in the liver is probably to supply precursors for various biosyntheses rather than to supply pyruvate for oxidation. Most of the TCA enzymes have been found in fish tissues and intermediates in the TCA cycle have been recovered, which indicates that this cycle is functional in fish. Enzymes for gluconeogenesis, the synthesis of glucose from other nutrients, are found in fish. Several studies have found high liver glycogen stores in starved fish, which suggest that gluconeogenesis, probably from amino acids, preempts glycogen breakdown for glucose production.

Metabolism of Lipids

Lipogenesis in fish proceeds similar to that in mammals. Fatty acids are synthesized through acetyl-CoA from 2-carbon residues that come primarily from glucose and deaminated amino acids. The 2-carbon units (acetate) are synthesized into primarily palmitic acid (C 16:0), but to a lesser degree into stearic (C 18:0) and myristic (C 14:0) acids. Generally, the scheme is as follows: acetyl-CoA is converted to malonyl-CoA by addition of CO_2; malonyl-CoA combines with a carrier protein (ACP) to form malonyl-ACP; malonyl-ACP combines with acetyl-ACP, releasing CO_2 and ACP, to form acetoacetyl-ACP; acetoacetyl-ACP is reduced, dehydrated and again reduced to form butyl-ACP, which is sequentially elongated with malonyl CoA units until palmitate is formed. The enzyme systems responsible for fatty acid synthesis are acetyl-CoA carboxylase and the fatty acid synthetase complex. Palmitate can be elongated and the fatty acids can be desaturated (double bonds produced) by enzymes in the mitochondria. In mammals, chain elongation does not go beyond C 18 and desaturation is limited to the omega-9 (n-9) position. Some fish species, such as rainbow trout, can chain elongate beyond C 18 and some can desaturate at n-6 as well as at n-9. Most of the polyunsaturated, long-chain fatty acids found in fish tissues are of dietary origin and come originally from microscopic aquatic plants.

Three synthesized fatty acids, or fatty acyl-CoA's, are subsequently esterified with glycerol-1-phosphate (which can come from the glycolytic pathway through dihydroxyacetone-phosphate) to form triglycerides. Liver and adipose tissues are primary sites for fatty acid synthesis.

Catabolism of fats in fish seems to be similar to the classical scheme identified in mammals. After hydrolysis, glycerol is phosphorylated and converted to dihydroxyacetone-phosphate, which enters the Embden Meyerhof glycolytic pathway where it can go to glucose or to pyruvate and the TCA cycle. The fatty acid is first activated to fatty acid acyl-CoA, which combines with carnitine to diffuse into the mitochondrion for oxidation. The carnitine diffuses back to the cytosol and the fatty acid acyl-CoA is degraded, two carbons at a time, which go into the TCA cycle as acetyl-CoA. This process, known as beta-oxidation, proceeds until all carbons are removed.

Except for a short time following a meal, most of the fat oxidized for energy comes from body fat stores. Release of triglycerides from adipose tissue is under hormonal control. Hydrolysis of fat to glycerol and fatty acids occurs in adipose tissue. The fatty acids circulate in the blood complexed with a protein. During starvation, mammals often develop "ketosis," which results from more acetyl-CoA being presented to the TCA cycle than there is oxaloacetate to carry it into the cycle. The surplus acetyl-CoA is converted to ketones, which produce an acidic condition in the blood. Teleost fishes, however, have been found not to produce significant amounts of ketone bodies. This may indicate that they are more efficient than mammals in oxidizing fatty acids for energy.

Metabolism of Amino Acids

The liver is responsible for maintaining the body amino acid pool. Sources of these amino acids are the diet, which is usually intermittent, and catabolized body proteins, which are in a continuous state of flux. In order to keep the pool relatively constant, amino acids are discretely released for synthetic purposes (proteins, nucleotides, etc.), for oxidation, for energy release, or for conversion to fat.

Protein Synthesis.   The mechanics of protein synthesis in animals was discussed in chapter 2. Rate of protein synthesis varies among fishes but is generally lower than in land animals. With intensive feeding, a broiler chicken will grow from 40 g to 1.8 kg, or increase its weight 45 times, in 7 to 8 weeks, whereas a young channel catfish (20 g) with intensive feeding will increase its weight only four to eight times during this time interval. Information on protein synthesis rate in fish is scarce, but in rats about 700 mg of muscle protein/100 g weight/day is synthesized. However, only about 150 mg of new tissue is gained, which indicates the high rate of protein turnover. The K_m constant for enzymes initiating protein synthesis in mammalian tissue is much lower than the K_m constant for enzymes degrading proteins and amino acids, meaning that protein synthesis can occur at very low amino acid concentrations, but that amino acid concentration must be relatively high for catabolism.

Amino Acid Degradation and Nitrogen Excretion.   Amino acids are not stored as such in large quantity in the body, as are fats or carbohydrates, so excesses are deaminated and the carbon residues oxidized or converted to fats, carbohydrates, or other compounds. The amino group is removed from amino acids mainly by transamination or by oxidative deamination. Transamination, which seems to be the major initial deamination pathway in fish, involves the transfer of ammonia from an amino acid to an α-keto acid, usually α-ketoglutarate. Aminotransferases are specific for the amino acid, and vitamin B_6 is a cofactor. The recipient alpha-keto acid releases ammonia through deamination for excretion or synthesis of other amino acids, and is recycled through another transamination. The keto acid formed in the initial transamination can be oxidized, converted to fat, or used in synthesis of other compounds. Of the essential amino acids, only lysine and threonine do not participate in transamination. The carbon skeleton of the amino acid is the part that determines whether or not an animal can synthesize it. The alpha-keto acids of all essential amino acids except lysine and threonine can be aminated metabolically and serve as essential amino acid sources.

Oxidative deamination is sometimes the initial deamination reaction. It is an energy yielding reaction catalyzed by dehydrogenase enzymes. The end products are an alpha-keto acid and free ammonia, because the amino group from the amino acid is not transferred to a keto acid.

Most terrestrial vertebrates excrete nitrogen in the urine as urea; however, birds and reptiles excrete uric acid, which requires less water. Teleost fishes excrete 60% to 90% of their waste nitrogen as ammonia, and most is eliminated through the gills. Other nitrogen excretory compounds for fishes are urea, uric acid, creatine, createnine, trimethylamine oxide, and amino acids. The ornithine-citrulline-arginine cycle does not seem to be the pathway for urea synthesis in fishes as it is in mammals. Some species, including the sharks, maintain high urea levels in the body fluids for osmoregulatory purposes.

Ammonia is an efficient route for nitrogen excretion. Apparently, less energy is expended than in the synthesis and excretion of urea. Also, non-ionized ammonia is lipid soluble and moves across cell membranes easily, and does not cause a loss of body water as in elimination of urea through the urine. Non-ionized ammonia is a relatively strong base and is highly toxic unless eliminated quickly, whereas urea is a weak base and is less toxic when it accumulates in the animal. Urea synthesis is an energy demanding (endergonic) process; ATP is required in the synthesis. Also, water is needed in urea excretion, which would be a liability for saltwater fishes, who need to conserve body water.

Most of the ammonia in the body fluids of fishes is in the ionized (less toxic) form, NH_4^+. At body pH of near 7, the ratio of NH_3 to NH_4^+ is about 1/100. (The pKa of ammonia, or pH at which NH_3/NH_4^+ is 1/1, is 9.24.) To cross cell membranes, ammonia must be in its non-ionized, lipid-soluble form. Although only 1% of the ammonia is in the non-ionized form at the surface of the cell membrane, this is not a real problem because of the rapid conversion of NH_4^+ to NH_3, which can diffuse from the extra-cellular body fluid into the gill and from the gill epithelial cells into the water. Goldstein, Forster, and Fanelli (1964) found that approximately 60% of the ammonia excreted from the gill of *Myoxocephalus scorpius* was from blood ammonia, while 40% was from plasma amino acids that were deaminated in the gill tissue.

The problem in ammonia excretion in fish is in the movement of ammonia from the gill to the water outside the fish's body. If ammonia concentration and pH of the water are low in relation to the fish's body fluid, NH_3 diffuses rapidly from the gill into the water. Once NH_3 is across the gill membrane and into the water, it changes to NH_4^+, the rate being dependent upon water pH. As water pH increases, the concentration of NH_3 in relation to NH_4^+ increases and makes movement of NH_3 from the gill epithelium more difficult. In fact, as concentration of NH_3 in water increases, the outward flow of NH_3 from the gill epithelial cells can be reversed.

Rate of ammonia production in fish culture systems is dependent on amount and quality (amino acid composition) or protein fed. The amino acids of ingested protein not used in body synthesis are deaminated and the nitrogen is excreted. If a protein of poor amino acid balance is fed, less protein synthesis per unit of ingested protein occurs and the unused amino acids are deaminated and the nitrogen excreted. Ammonia production by a 1-kg channel catfish fed a 32% protein commercial feed is approximately 600 mg of ammonia per kilogram of fish weight per day. This is based on the following assumptions: The fish consumes 25 g of feed/kg of fish weight/day; 20% of the fed protein is undigested, 40% is retained by the fish, and the remaining 40% is excreted by the fish as ammonia or products that will rapidly be converted to ammonia by bacteria. Calculation is as follows:

$$(25 \text{ g food}) \times (32\% \text{ protein}) \times$$

$$(16\% \text{ N in protein}) \times (40\% \text{ of N excreted}) \times \qquad (3.3)$$

$$(1.2 \text{ g } NH_3 \text{ per g of N}) = 610 \text{ mg ammonia.}$$

The lethal level of non-ionized ammonia for channel catfish is around 2.5 mg/L of water. Sublethal levels causing gill tissue damage in channel catfish and rainbow trout are 0.12 mg/L to 0.4 mg/L.

RATE OF METABOLISM (OXYGEN CONSUMPTION) IN FISH

Rate of metabolism, or oxygen consumption, in fish is influenced primarily by water temperature, fish age or size, activity of the fish, feeding, and concentration of oxygen in the water. Concentration of ammonia, nitrite, or other adverse compounds in the water, nutritional deficiencies, hormone applications, and other factors can also affect rate of metabolism in fish. Unlike in homeothermic animals, metabolism in fish is directly related to temperature. Thus, fish will respond more slowly to stress or nutrient deficiency at lower water temperatures.

Metabolic rate in mammals is usually reported as heat production (kcal/kg body weight/hr). This can be determined by placing the animal in a calorimeter and measuring heat production directly or by

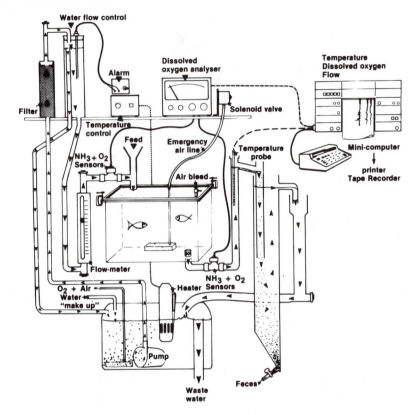

Fig. 3.6.   Computer controlled fish respirometer equipped with a fish tank, feces settling column, and instrument to measure oxygen consumption and water temperature (Cho et al. 1982).

measuring oxygen consumption and calculating heat production from the calorie equivalent value of the oxygen consumed (4.6–5.0 kcal/L of oxygen). Oxygen consumption is the preferred method for measuring and reporting metabolic rate in fish. A number of respirometers have been designed and used successfully to measure oxygen consumption by fish under various conditions of activity, feeding, and temperature. Figure 3.6 shows a fish respirometer designed by Cho, Slinger, and Bayley (1982). If oxygen consumption is to be converted to heat production in fish, oxygen caloric equivalent values of 4.63 kcal/L of O_2 consumed should be used when the energy sources are protein and fat and 4.8 kcal/L of O_2 consumed when some carbohydrate is being oxidized.

Knowledge of metabolic rate of fish under various conditions has practical importance. It is valuable in determining dietary energy levels and daily feed allowances for fish under various conditions and for various production purposes. For example, maintenance energy requirements of fish and heat increment (loss) values for various feedstuffs can be derived from metabolic rate data. Oxygen consumption rate by fish is also useful in determining carrying capacity of fish culture systems and in predicting aeration needs or flow rates in various aquacultural environments.

Effects of Temperature

Temperature has little effect on metabolic rate in warmblooded animals unless it is above or below the "comfort range" of the animal. In fishes, metabolic rate varies directly with temperature. Within the normal environmental temperature range of the fish, a Q_{10} value of 2.3 has been found to apply within a 20% span for nonfed brook trout, bullhead catfish, carp, and sockeye salmon (Brett 1979). Q_{10} represents the increase in metabolic rate for an increase of 10° C. Andrews and Matsuda (1975) found Q_{10} values of 2.3 and 1.9 for nonfed and fed channel catfish, respectively.

Effects of Fish Size

Andrews and Matsuda (1975) measured oxygen comsumption by channel catfish ranging from 2.5 g to 1,000 g under feeding and nonfeeding conditions. As shown in Table 3.1, oxygen consumption rate decreased markedly as fish size increased.

Table 3.1. OXYGEN CONSUMPTION RATE BY NONFED AND FED CHANNEL CATFISH OF FOUR SIZES MEASURED AT 26° C

Fish size (g)	Oxygen consumption rate (mg/kg/hr)		Increase in oxygen consumption from feeding (%)
	Nonfed[1]	Fed[1]	
2.5	880	1,230	40
100	400	620	55
500	320	440	38
1,000	250	400	60

[1] Nonfed fish were fasted 24 hr; fed fish were fed to satiation on a commercial feed 1 hr before measurement.
Source: Andrews and Matsuda (1975).

Effects of Feeding

Following ingestion of food, an increase in metabolic rate occurs in animals. This is known as the heat increment. In mammals, it can amount to up to 30% of the caloric content of the diet. In fish it is generally lower, usually 10% to 15%. This increase in metabolism is due primarily to metabolism of amino acids. A number of studies with fish and warmblooded animals have shown that heat increment is proportional to percentage of protein in the diet. Ingestion of carbohydrates and lipids will also elicit an increase in heat increment, but less than that from protein. The increase in oxygen consumption subsequent to feeding is large. As shown in Table 3.1, channel catfish increased oxygen consumption within 1 hr after feeding by 38% to 60%. Duration of heat increment varies with fish species, size and compositon of meal consumed, and environmental conditions. Typically, when fish are fed once daily, heat increment begins immediately after feeding, peaks after 4 to 8 hours, and terminates within 18 to 24 hours.

REFERENCES

ANDREWS, J. W., and Y. MATSUDA. 1975. The influence of various culture conditions on oxygen consumption of channel catfish. *Trans. Amer. Fish. Soc.* 104: 322–327.

BRETT, J. R. 1979. Physiological energetics. In *Fish Physiology,* vol. VII, ed. W. S. Hoard, D. J. Randall, and J. R. Brett. New York: Academic Press.

CHO, C. Y. 1982. Effect of dietary protein and fat levels on heat increment of feeding rainbow trout. *Proc. XII Int. Cong. Nutr.,* San Diego, CA.

CHO, C. Y., S. J. SLINGER, and H. S. BAYLEY. 1982. Bioenergetics of salmonid fishes: Energy intake, expediture and productivity. *J. Biochem. Physiol.* 73B: 25–41.

CHOUBERT, G., JR., J. DE LA NOUE, and P. LUQUET. 1979. Continuous quantitative collector for fish feces. *Prog. Fish.-Cult.* 41: 64–66.

GOLDSTEIN, L., R. P. FORSTER, and G. M. FANELLI. 1964. Gill blood flow and ammonia excretion in *Myxocephalus scorpius. Comp. Biochem. Physiol.* 12: 489–499.

GRIZZLE, J. M., and W. A. ROGERS. 1976. *Anatomy and histology of channel catfish.* Auburn, AL: Agricultural Experiment Station, Auburn University.

POPMA, T. J. 1982. Digestibility of selected feedstuffs and naturally occurring algae by tilapia. Ph.D. diss., Auburn University, Auburn, AL.

SMITH, B. W., and R. T. LOVELL. 1971. Digestibility of nutrients in semipurified diets by channel catfish in stainless steel troughs. *Proc. 25th Ann. Conf. of S.E. Assoc. Game and Fish Comm.* 452–459.

SMITH, B. W., and R. T. LOVELL. 1973. Determination of apparent protein digestibility in feeds for channel catfish. *Trans. Amer. Fish. Soc.* 102: 831–835.

4

Nonnutrient Diet Components

In addition to the essential nutrients, feeds may contain organic and inorganic materials that have beneficial, negligible, or deleterious effects on the growth or health of the fish or the sensory quality of the processed fish. These may be naturally occurring, intentionally or unintentionally added, or products produced through chemical change or microbial growth.

TOXINS AND ANTIMETABOLITES

Microbial Toxins

The microbial toxins of greatest economic importance in animal feeding are mycotoxins—metabolites of toxigenic molds. Those of current concern in the United States are produced by certain species of three genera of molds: *Aspergillus, Penicillium,* and *Fusarium.* These molds are ubiquitous and grow and produce toxins under conditions that include adequate substrate (carbohydrate), minimum moisture in the substrate of 14%, relative humidity of 70% or above, adequate temperature (varies widely with different molds), and oxygen. Mycotoxins are usually produced in feedstuffs prior to harvest, but can be produced in the feedstuffs or finished feeds during improper storage.

Aflatoxin. Aflatoxins, produced by *Aspergillus* sp., are the mycotoxins of greatest concern in animal feeding in the United States because of their toxic and carcinogenic properties and frequent occurrence, especially in the Southeast. Common effects of aflatoxin consumption among farm animals are poor growth, liver damage, impaired blood clotting, decreased immune responsiveness, and increased mortality.

Rainbow trout has one of the highest sensitivities to aflatoxin of all animals (Halver and Mitchell 1967). Less than 1 μg/kg of diet will cause liver tumors (see Figure 4.1). The LD$_{50}$ (dose causing death in

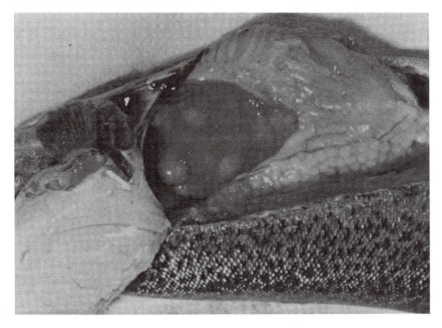

Fig. 4.1.   Aflatoxin-induced hepatoma in rainbow trout. The fish was fed 1 mg of aflatoxin per kg of diet for 20 wk.
(From Halver and Mitchell 1967)

50% of the subjects) for aflatoxin in 50-g rainbow trout is 0.5 mg/kg to 1.0 mg/kg of diet. Signs of severe aflatoxicosis in rainbow trout are liver damage, pale gills, and reduced red blood cell concentration. Warmwater fish are less sensitive to aflatoxin. The LD_{50} for channel catfish is approximately 30 times that for rainbow trout. Pathological signs in channel catfish fed lethal doses of aflatoxin are death and lesions in liver, lining of the stomach, intestines, spleen, heart, and kidney.

Traditionally, feed sources in the United States most likely to be contaminated with aflatoxin are corn, cottonseed, and peanuts. Corn from the Southeast is more likely to be contaminated with aflatoxin than that from the Midwest due to climatic conditions. In 1977, a bad year for aflatoxin contamination in corn, 1,868 corn samples were collected from states in the Southeast and 27% contained over 400 μg/kg, 24% contained 100 μg/kg to 400 μg/kg, and only 25% contained less than 20 μg/kg, which is the maximum level allowed by the U.S. Food and Drug Administration. Normally, only sporadic aflatoxin problems are found in corn from the Midwest. Peanuts grown in the Southeast vary in aflatoxin contamination from year to year. Cotton from the Southeast rarely contains aflatoxin; that which does is generally from the arid, irrigated regions of the West.

Fusarium. The *Fusarium* toxins that are most harmful to animal health are zearalenones and tricothecenes. The zearalenones are a group of estrogenic metabolites, some of which cause reproductive problems in farm animals consuming 0.6 mg/kg to 5 mg/kg in diets. The tricothecenes usually develop in corn in storage in the Midwest, with alternating cooling and warming trends in the fall. Toxicity signs in livestock and poultry are reduced growth, reduced red blood cell formation, widespread hemorrhage, poor blood clotting, impaired immune responses, and death. *Fusarium* toxicity signs have not been described in fish.

Ochratoxins. Ochratoxins are produced by *Aspergillus* and *Penicillium* mold species that are widely found in nature. These toxins have not been recognized as causing widespread problems in animal feeding in the United States. It is suspected, however, that poor growth and feed conversion in livestock occur in undetected cases from ochratoxin contamination. Ochratoxins are recognized as kidney toxins, causing pale, swollen kidneys and renal tubular failure in swine, rats, and mice. The LD_{50} for ochratoxin A in 6-month-old rainbow trout is 4.7 mg/kg of diet. Pathological signs in trout fed ochratoxin A have been identified as severe necrosis of liver and kidney tissues, pale kidneys, pale, swollen livers, and death.

Other Microbial Toxins. There are other mold toxins, such as cyclopiazonic acid, vomitoxin, slaframine, and citrinin, whose pathologies have been demonstrated in warmblooded animals but not in fish. Some molds and bacteria can destroy nutrients in feeds. An example is *Pseudomonas* species of bacteria and *Aspergillus* species of molds, which can separate glutamic acid from pteroic acid in the vitamin, folic acid, thus causing folic acid deficiency.

Microbial Toxins in Commercial Fish Feeds. Aflatoxicosis has caused considerable economic losses in livestock. This has been primarily where the feeds were fed to livestock on the farm where they were produced and the feeder had no knowledge of aflatoxin contamination. Commercially processed feeds are less likely to contain mycotoxins because of government regulation on tolerance levels of aflatoxins in feedstuffs sold in interstate commerce, screening of feeds for aflatoxins by the feed manufacturer, and the fact that a contaminated feedstuff is usually blended with noncontaminated feedstuffs, which dilutes the toxin.

Aflatoxin in cottonseed meal was responsible for serious economic losses in hatchery-raised trout in the 1960s until research revealed the intense sensitivity of rainbow trout to the toxin. Mycotoxins have not been conclusively identified as problem contaminants in catfish feeds, although situations have occurred where mycotoxicosis was suspected. Catfish feeds often contain grains produced in the southeastern United States, which sometimes contain high concentrations of aflatoxin. Aflatoxin is relatively heat stable and little is destroyed in feed processing. Catfish feeds containing 100 μg/kg to 200 μg/kg of aflatoxin have been reported. It is conceivable that such feeds could affect fish performance because aflatoxin levels of 200 μg/kg of diet reduce weight gain, impair blood coagulation, and decrease immune responses in pigs, cattle, and turkeys.

Control of Microbial Toxins in Fish Feeds. Although the Food and Drug Administration (FDA) requires that grains shipped interstate contain less than 20 ppb of aflatoxin, grains produced and sold within the state are not subject to FDA regulation. Thus, feed manufacturers should require all feedstuffs that are traditionally associated with mycotoxins be tested before using. Feed manufacturers should avoid using feedstuffs that are suspected of having any trace of aflatoxin in feeds for fry fish because of the increased sensitivity of young fish.

Feed processors should be reminded that steam pelleting or even extrusion processing of feeds does not destroy aflatoxin, because it is very stable to heat. Even though most mycotoxins are produced in the ingredient before the feed is processed, aflatoxins also can develop in fish feeds after processing if they are poorly dried or stored under highly humid conditions. The use of an anti-mold additive in the feed will minimize this possibility.

Histamine and Gizzerosine

These are moderately toxic products that have been found in fish meals. Histamine is a ptomaine, produced from bacterial and autolytic decarboxylation of the amino acid histidine. This compound is produced during decomposition of improperly stored fish prior to reduction to fish meal. It has been found in higher concentrations in meal from dark- than from light-fleshed fish. Reduction in growth rate has been reported in fish fed diets containing high concentrations of histamine.

Gizzerosine [2-amino-9-(4-imidazolyl)-y-azanonanoic acid] was recently isolated from fish meal and identified as a causative compound for gizzard erosion in broiler chickens fed diets containing heavily heated fish meals (Okayaki et al. 1983). Gizzerosine is produced by the reaction between free histidine and some side chain in the protein; the reaction is enhanced by heating. The toxic substance is in the protein and not as a free histidine derivative. Gizzerosine and histamine both derive from the amino acid histidine, but the former appears to result from a heat induced reaction, whereas the latter is produced from autolytic or microbial enzyme action. Gizzerosine has not been investigated as a toxicant in fish feeds, but there is a possibility that it could be.

Phytic Acid

Phytic acid, found in most plant feedstuffs, is a hexaphosphoric acid ester of inositol. It occurs as salts of calcium, magnesium, and other divalent cations. Approximately 60% to 70% of the phosphorus in plant feedstuffs is in phytic acid and is poorly available to fish. This is well known and is usually considered when determining phosphorus allowances for fish feeds. Phytic acid reduces bioavailability of zinc and other elements. Gatlin and Wilson (1984) showed that channel catfish fed a soybean meal based diet, containing 1.1% phytic acid, required supplementation with 100 mg of zinc above the normal requirement of 20 mg/kg of diet. Phytic acid also depresses protein digestibility, as has been demonstrated with rainbow trout.

Gossypol and Cyclopropionic Acid

Gossypol is found in the pigment glands in most commercial cotton varieties. Free gossypol is moderately toxic to nonruminant animals and limits the quantity of cottonseed meal that can be used in swine, poultry, and fish feeds. A dietary level of 0.03% free gossypol suppresses growth rate and a level as low as 0.01% will cause liver damage in rainbow trout (Herman 1970). A dietary level of 20% or more of prepress-solvent extracted cottonseed meal in channel catfish diets causes gossypol toxicity, but lower levels do not (Robinette 1981).

The amount of free gossypol in cottonseed meal depends upon the methods of processing the seed. Generally, free gossypol contents of three types of cottonseed meal are as follows (Watts 1970): direct solvent extracted meal, 0.2% to 0.4%; screwpress meal, 0.02%; prepress solvent extracted, 0.05%. If the maximum level of free gossypol for fish diets is 0.03%, then 7.5% to 15% would be the highest level of direct solvent processed cottonseed meal to use in the diet. Higher levels of the other two meals could be used. The addition of 0.85 to 1.0 part ferrous sulfate to each part free gossypol in diets of swine (Tanksley 1970) and poultry (Watts 1970) has proven successful in blocking the toxic effects of gossypol. This has not been tested in fish.

All varieties of cottonseeds contain cyclopropenoic fatty acids that produce several undesirable effects in birds, mammals, and fish. Increased saturated fatty acids in body lipids and delayed sexual maturity occurred in female rats; increased cholesterol levels, aortic atherosclerosis, and liver damage was found in rabbits. Liver damage, increased glycogen deposition, and elevated saturated fatty acid levels in lipids resulted when rainbow trout were fed cyclopropenoic fatty acids.

Oxidized Fish Oil

Marine fish oils contain 20% to 25% polyunsaturated fatty acids. Autoxidation of unsaturated fatty acids produces a large number of free radicals and peroxide compounds, which are active pro-oxidants. These components may react with other diet ingredients and reduce their nutritional value or, after ingestion, react with oxidation-sensitive phospholipid cellular and subcellular membranes and cause damage. Ingestion of oxidized fish oils has caused reduced growth rate, anemia, nutritional muscular dystrophy, and lesions and ceroidosis in liver of fishes (see Figure 4.2). Increasing dietary levels of vitamin E reduces severity of these toxicity effects.

Fiber

Fiber is not a specific chemical compound, but a mixture of lignin, cellulose, hemicellulose, pentosans, and other components that are generally indigestible to monogastric animals, including fish. Fiber has no functional value in fish feeds except possibly to control rate of movement of ingesta through the digestive tract, and this is probably unnecessary in practical fish diets. Dupree and Sneed (1966) increased growth rate in channel catfish by increasing fiber content from 0% to 21% in purified diets. The fiber probably reduced the passage rate of the diet in the digestive tract and increased digestion. However, Leary and Lovell (1975) found no benefit in increasing fiber in practical catfish diets above the basal content of 2.8% and that increasing fiber beyond 14% reduced growth rate.

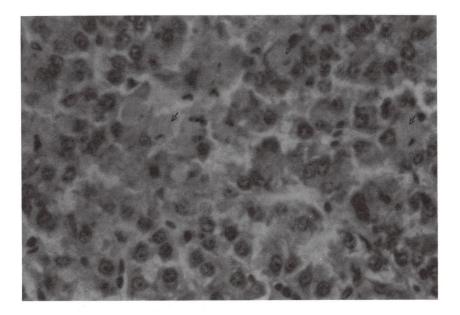

Fig. 4.2.   Accumulation of ceroid (arrows) in macrophages in liver of rainbow trout fed oxidized fish oil.
(Courtesy of Charlie E. Smith)

Antiproteases

Nonheated legume seeds, especially soybeans, contain globulin protein that combines with and inactivates pancreatic and intestinal trypsin and chymotrypsin. This significantly reduces growth rate in monogastric animals. In addition to reduction in protein digestion, hypertrophy and excess secretion from the pancreas have been observed in rats fed either nonheated soybeans or isolated protease inhibitor. Fish do not utilize nonheated soybean products well. In fact, additional heating of commercially processed soybean meal has reduced activity of the protease inhibitors and increased growth rate of rainbow trout and channel catfish over that obtained from feeding the commercial meal without further heating.

Thiaminases

Tissues of most freshwater fishes and some saltwater species contain an enzyme that can hydrolyze thiamine. The enzyme is heat sensitive and can be inactivated by moderate heating of the fish flesh before feeding. Thiamine in prepared fish diets is only destroyed after contact with the thiaminase for a period before being consumed. Raw fish products can be fed to fish if a source of thiamine is fed in a separate diet. Otherwise, the fish products should be heat treated before feeding.

DIET ADDITIVES

Hormones

Hormonal control of sexual development is advantageous in species where it is desirous to grow monosex (one sex) fish in the culture system to prevent reproduction or increase growth rate. Incorporating androgenic steroids (ethyltestosterone and methyltestosterone) in diets (30 to 60 mg/kg) fed as first food to tilapia fry (*Tilapia mossambica, Oreochromis nilotica,* or *O. aurea*) and continued for 14 to 21 days results in 90% to 100% development of male fish. This method of sex reversal in tilapia is practiced commercially in several countries; however, feeding steroids for sex control is not approved for use in the United States. Experimental feeding of lower doses of 17-β-methyltestosterone to rainbow trout and Atlantic salmon fry for three months after the fish begin feeding resulted in all male fish. Feeding androgenic steroids to channel catfish fry, however, resulted in female development in all fed fish. Goudie et al. (1983) theorized that the fed androgen (17-β-methyltestosterone) was enzymically changed to a compound with estrogenic properties in the fish. Feeding estrogenic steroids to tilapia fry (ethynylestradiol, estrone, diethylstilbestrol) and salmonid fry (17-β-estradiol) caused development of all females.

Experimental feeding of low levels (1.0–2.5 mg/kg of diet) of 17-α-methyltestosterone markedly improved growth and feed efficiency of juvenile coho salmon, chinook salmon, and rainbow trout. Other steroids that have increased growth rate of salmonids at low feed levels include 11-ketotestosterone, oxymethalone, and testosterone. Low dose feeding of androgens also increased growth rate in carp and eels, but not in channel catfish. Estrogenic hormones have not been as effective for enhancing growth as androgenic hormones in salmonids. Steroid hormones fed to salmonids at levels exceeding 10 mg/kg of diet or to channel catfish at levels as low as 2 mg/kg resulted in growth suppression, gonadal tissue degradation, and skeletal deformities (see Figure 4.3). Although some synthetic hormones are approved for use in the United States to improve growth and feed efficiency with cattle, primarily as implants, such compounds have not been cleared by the Food and Drug Administration for fish.

Pellet Binders

Steam pelleted fish feeds, especially those fed to crustaceans, usually contain added binders to improve water stability. A list and discussion of binding agents used in laboratory and commercial diets are presented in chapter 5 on Feed Formulation and Processing. There is no

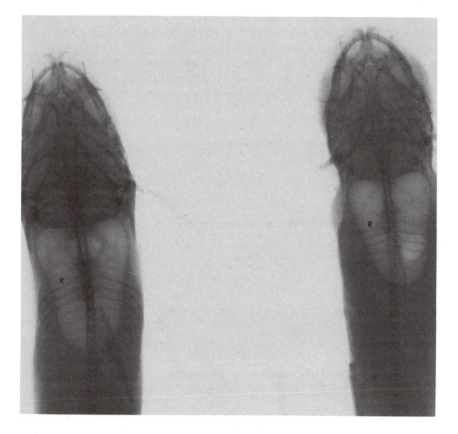

Fig. 4.3. Feeding 2.5 mg of 17-methyltestosterone per kg of diet impaired bone development in channel catfish on the right; fish on left is control. The bones were weaker and malformed in the hormone-fed fish. Note smaller rib bones in the fish on the right.
(Courtesy of A. L. Gannam)

evidence that the commonly used organic binders, which are lignosul-
fonates of hydrocolloidal polysaccharides or their derivatives, have
detrimental effects on fish. The polymethylocarbamide *Basfin* (BASF,
West Germany) caused reduction in growth and food consumption in
some species when added to diets at levels of 0.75% or above.

Antibiotics

Three antibiotics are currently registered in the United States for
feeding fish for disease control, as shown in Table 4.1. Antibiotics are
used at subtherapeutic levels in poultry and livestock feeds for growth
promotion. This has not been found advantageous in studies where
oxytetracycline and sulfonamides were fed to various salmonids.
Sulfonamids fed to cutthroat trout brood fish at the therapeutic dose
level resulted in kidney histopathology and increased mortality.
However, subtherapeutic levels (0.01%) of furazolidone stimulated
growth and feed efficiency in red sea bream (Yone 1968).

Table 4.1. ANTIBIOTICS LICENSED FOR USE IN FEEDS FOR FOOD FISH IN THE
UNITED STATES

Product	Use	Amount to feed
Romet-30 (Sulfadimethoxine + ormetoprim)	Antibacterial against furunculosis in salmonids and *Edwardsiella ictaluri* in channel catfish. Withdrawal period prior to slaughter is 21 days for salmonids and 3 days for channel catfish.	50 mg/kg of fish per day for 5 days
Sulfamerazine, for fish	Antibacterial against furunculosis in salmonids. Do not feed within 21 days of slaughter.	22 mg/100 g of fish per day for 14 days
Terramycin, for fish	Antibacterial against *Aeromonas, Hemophilus,* and *Pseudomonas* in fish. Do not feed within 21 days of slaughter.	5.5–8.25 mg/100 g of fish per day for 10 days

Attractants

Some materials, with or without nutritive value, are added to fish feeds to serve as attractants or to enhance palatability. Squid meal, shrimp head meal or extracts have been reported to improve acceptance of feed to some penaeid shrimps. Fish meal is considered an attractant in many fish feeds. Some chemical compounds, such as free amino acids, are highly effective olfactory and gustatory stimuli for fishes, but whether amino acids like alanine and arginine, which are highly taste effective to channel catfish, signal a food source in the environment is not determined. Other soluble organic compounds have been shown to elicit feeding responses or preferences from fish in aquaria. The efficacy of using such compounds in practical feeding should be verified.

Antioxidants

Oxidation of lipids in feeds or feedstuffs can cause reduction of the nutritional value of certain lipids and vitamins. This can also produce pro-oxidative compounds that are toxic to the fish, expecially if vitamin E or selenium is marginal or deficient in the diet. To preserve oxidation-sensitive nutrients and prevent formation of toxic peroxide compounds, synthetic antioxidants or additional levels of natural antioxidants like vitamin E should be included in fish feeds. Synthetic antioxidants such as BHA (butylated hydroxyanisole), BHT (butylated hydroxytoluene), and ethoxyquin (1,2-dihydro-6-ethoxy-2,2,4-trimethylquinoline) are effective synthetic antioxidants commonly used in animal feeds. Maximum levels permitted by the United States Food and Drug Administration is 0.02% of the fat content for BHA and BHT and 150 mg/kg feed for ethoxyquin.

ACCIDENTAL CONTAMINANTS

These include elements or compounds such as heavy metals, pesticides, and industrial chemicals that enter the feed or its ingredients unintentionally or by accident during production, processing, or storage. Examples that have been identified are fish meal containing mercury accumulated from the sea, plant products containing pesticide residues from excessive farm application, overfortification of trace mineral or medicinal supplements in the feed, and contamination of the feed from

Table 4.2. FOOD AND DRUG ADMINISTRA-
TION ACTION LEVELS FOR TOXIC
OR DELETERIOUS SUBSTANCES
IN FINISHED ANIMAL FEEDS

Substance	Quantity mg/kg
Aflatoxin	0.02
Aldrin and Dieldrin	0.03
Benzene Hexachloride (BHC)	0.10
Chlorodane	0.10
Dibromochloropropane (DBCP)	0.05
DDT, DDE, and TDE	0.50
Endrin	0.03
Heptachlor and Heptachlor Epoxide	0.03
Kelthane	0.50
Lindane	0.10
Polybrominated Biphenyls (PBB's)	0.05
Toxaphene	0.50

Sources: All substances except PBB's: Food and
Drug Administration Compliance Policy Guides,
Guide 7126.27, Animal Feeds—chapter 26,
1980. For PBB's: Congressman William M.
Broadhead's petition to reduce FDA action levels
for PBB's in food, July 27, 1977.

pesticide applications during storage. These are highly uncommon, but
have occurred and caused significant stress or loss of fish.

Feed ingredients may occasionally exceed FDA tolerance levels for
certain pesticides. The feed manufacturer should be aware of these
tolerances (Table 4.2) and have all ingredients with high lipid concentrations monitored periodically for pesticide residues. Most pesticides
are lipid soluble and are found in lipid fractions of the feedstuff.

REFERENCES

DUPREE, H. K., and K. E. SNEED. 1966. Response of channel catfish fingerlings to
various levels of major nutrients in purified diets. *U.S. Bur. Sport Fish. Wildl. Tech.
Pap.* 9: 1–21.

GATLIN, D. M. III, and R. P. WILSON. 1984. Zinc supplementation of practical
catfish diets. *Aquac.* 41: 31–36.

GOUDIE, C. A., B. D. REDNER, B. A. SIMCA, and K. B. KAVIS. 1983. Feminization
of channel catfish by oral administration of steroid sex hormones. *Trans. Am. Fish.
Soc.* 112: 670–672.

HALVER, J. E., and I. A. MITCHELL. 1967. Trout hepatoma research conference
papers. *Bur. Sport Fish. and Wildlife Rep.* 70. Washington, DC: U.S. Department of
the Interior.

HERMAN, R. L.   1970.   Effects of gossypol on rainbow trout *Salmo gairdneri*. *J. Fish. Biol*. 2: 293–303.

LEARY, D. F., and R. T. LOVELL.   1975.   Value of fiber in production diets for channel catfish. *Trans. Am. Fish. Soc*. 104: 328–332.

OKAYAKI, T., T. NOGUCHI, K. IGARASHI, Y. SAKAGAMI, and H. SETO. 1983.   Gizzerosine, a new toxic substance in fish meal, causes severe gizzard erosion in chicks. *Agric. Biol. Chem*. 47: 2949–2952.

ROBINETTE, H. R.   1981.   Use of cottonseed meal in catfish feeds. *Proc. Catfish Farmers of Amer. Res. Workshop* 3: 26.

TANKSLEY, T. D., JR.   1970.   Use of cottonseed meal in swine rations. *Feedstuffs* 42: 22–23.

WATTS, A. B.   1970.   Use of cottonseed meal in young chick rations. *Feedstuffs* 42: 23–24.

YONE, Y.   1968.   Effect of furazolidone on growth and feed efficiency of sea bream. *Bull. Jpn. Soc. Sci. Fish*. 34: 305–309.

5

Feed Formulation and Processing

FORMULATING FISH FEEDS

When fish are cultured in a system where natural foods are absent, such as trout raceways, or where natural foods make only a small contribution to the nutrition of the fish, as in intensively stocked catfish ponds, the feed should be nutritionally complete. However, where abundant natural food is available, supplemental feeds need not contain all of the essential nutrients. Nutritionally incomplete feeds are usually specific for a fish species or culture system; supplemental feeds for tilapia culture are discussed in chapter 8. This chapter discusses the formulation of nutritionally complete feeds.

Nutrient Requirements

The nutrient requirements presented in this book, NRC bulletins, and other publications were determined primarily with small fish and represent levels affecting maximum growth rate. Also, many of the requirements represent single laboratory experiments, and are unchallenged or unsupported by other studies. Thus, many requirements are, at best, only rough approximations of the optimum amounts of nutrients for practical diets to grow fish to harvestable size. Management, environmental factors, and fish size can have an effect on dietary nutrient levels for optimum performance. However, nutrient requirement data that are available serve relatively well as a basis to formulate highly productive, economical diets for commercial aquaculture. In formulating a diet for a species where nutrient requirements are not known, the requirements for a related species whose nutrient requirements are known can be discretely substituted. Generally, most variation among fishes should be expected between warm- and cold-water species, fresh- and saltwater species, and finfishes and crustaceans. As more information becomes available on nutrient requirements of various species, the recommended nutrient allowances for specific needs will become refined and practical feeds will be more cost effective.

Overfortification of labile nutrients in processed fish feeds is necessary. Values in nutrient requirement tables usually represent minimum requirements and do not allow for processing or storage losses. Amino acids and inorganic nutrients are relatively stable to heat, moisture, and oxidation, but many vitamins are not. Table 5.1 shows rates of loss of vitamins during steam pelleting and extrusion processing. Approximately 50% of the vitamin C is lost during extrusion and the half-life of vitamin C in fish feed after processing is approximately 2.5 to 3 months. Other vitamins are more stable than vitamin C, but losses still occur during processing and storage.

Physical Properties

Early experiments demonstrated that channel catfish and common carp utilized pelleted feed more efficiently than meal. Many of the small particles were not ingested, resulting in poorer feed conversion and reduced water quality in the ponds. To minimize these undesirable effects, commercial feeds should be processed into pellets that will remain intact in water until consumed by the fish. Pellets with long water stability are especially important for slow-feeding species such as shrimp.

Particle size of fish feeds should be as large as possible to minimize nutrient losses through leaching into the water, but not too large for the fish to consume. Optimum sizes of feed particles for channel catfish, trout, tilapia, and shrimp of various sizes are discussed in chapters on practical feeding of these species. Diet texture is important for some

Table 5.1. STABILITY OF ADDED VITAMINS IN EXTRUSION PROCESSED CATFISH FEED

Vitamin	Percentage recovery in extruded feed
Vitamin A acetate (in gelatin-starch beadlet)	65
Thiamin mononitrate	64
Pyridoxine hydrochloride	67
Folic acid	91
DL-alpha-tocopherol acetate	100
Ascorbic acid (ethylcellulose coated)	43

Source: Producer's Feed Company, Belzoni, MS; assayed by Hoffman-LaRoche, Inc., Nutley, NJ.

fishes. Most commercially cultured species accept dry feed particles, but some fish, such as eels and young salmon, prefer soft diets.

Catfish farmers in the United States prefer floating feeds. This is a valuable management tool when raising fish in ponds because it allows the fish farmer to determine how much the fish are consuming and also allows him or her to gauge the condition of the fish and water quality on the basis of feeding activity. Most fishes accept surface feeds satisfactorily, but some species, such as penaeid shrimps, prefer sinking feeds.

Computer Formulation of Fish Feeds

To determine minimum-cost feed formulations, the following information must be available:

1. cost of feed ingredients;
2. nutrient content of feed ingredients;
3. nutrient requirements of the animal;
4. availability of nutrients to the fish from various feed materials; and
5. minimum-maximum restrictions on levels of various ingredients.

Items 1 and 2 are readily available for most commercial feedstuffs. Enough information is available on nutrient requirements (item 3) for several fish and shellfish species to formulate satisfactory production diets for most commercially cultured fish. The availability of nutrients to fish from various types of feed materials must be known in order to make computerized substitutions among ingredients. Digestible or metabolizable energy and nutrient values for commercial ingredients are presented in Appendix A. These values are more limited and variable for fish than for livestock. For example: digestibility of energy from carbohydrates is much less for coldwater fish than for warmwater species; digestibility of phosphorus is less for fish than for livestock, especially for fish without gastric sections in the digestive tract; and the lysine in cottonseed meal is only 75% as digestible as the lysine in soybean meal.

Table 5.2 presents nutrient and ingredient restrictions that have been used for least-cost formulation of catfish feeds. Theoretically, protein level does not have to be restricted if essential amino acid requirements are well known or unless protein supplies another essential nutritional need, such as energy. Experience has shown that for channel catfish, if the minimum requirements for lysine and the sulfur amino acids (methionine and cystine) are met, the requirements

Table 5.2. NUTRIENT AND INGREDIENT RESTRICTIONS FOR LEAST-COST
FORMULATION OF A PRODUCTION FEED FOR CHANNEL CATFISH

Item	Restriction	Amount	Unit
Crude protein	Min.	32.0	%
Digestible energy	Min.	2.8	kcal/g
	Max.	3.0	kcal/g
Crude fiber	Max.	7.5	%
Available phosphorus	Min.	0.5	%
Digestible lysine	Min.	1.63	%
Digestible methionine + cystine	Min.	0.74	%
Whole fish meal or comparable animal protein	Min.	6.0	%
Cottonseed meal	Max.	10.0	%
Xanthophylls	Max.	11.0	mg/kg
Trace mineral mix	Include		
Vitamin mix	Include		

for the other eight essential amino acids will also be met. Other nutritional restrictions are minimum available phosphorus and digestible energy. Only maximum calcium is considered because dissolved calcium in the water will usually make a deficiency unlikely, but a mineral imbalance can result from high levels of bone ash (indicated by high calcium) in the feed. Whole fish or other animal protein sources have been found to be beneficial in catfish feeds for reasons not explained on the basis of meeting amino acid requirements. Cottonseed meal is restricted, because of free-gossypol toxicity to channel catfish. Carotenoid (xanthophyll) level is restricted, because it imparts undesirable yellow pigmentation to catfish flesh.

Limitations to computer formulation of fish feeds should be recognized. Some considerations may not be programmable because of milling characteristics. For example, sorghum does not extrude as well as corn in processing floating catfish feeds. Maximum advantage of least-cost formulation is realized when the processor has access to and is able to work with a variety of ingredients. In many cases logistics and handling facilities limit the number of ingredients with which the processor can work. As discussed previously, availability of nutrients among feedstuffs varies, but knowledge of the substitutability of one feedstuff for another is essential. Also, when formulating least-cost feeds, considerations other than nutritional must be made, such as the physical, palatability, and toxicological properties of the feed.

Use of quadratic growth responses to select protein or amino acid concentrations for production feeds is used to formulate poultry feeds. Arraes (1983) showed that feeds derived from quadratic programming constraints produced market-size broiler chickens for 5% to 15% less cost than linear programming that used NRC nutrient requirements. The NRC requirements are for a fixed, usually maximum, rate of growth. Quadratic programming takes into account diminishing productivity with increasing nutrient inputs. It also considers changes in value of growth produced as well as changes in cost of production.

In order to use quadratic programming, biological response functions from actual feeding trials are needed. With known growth responses from changing nutrient inputs, that is, regression of weight gain on nutrient concentration in the diet, the most profitable concentration of the nutrients for given cost-price conditions can be derived. Santiago (1985) presented response functions (quadratic regressions) for the ten essential amino acids for Nile tilapia, which is discussed in chapter 6. Much of the growth response data from feeding trials where fish have been fed various concentrations of protein, amino acids, or energy can be fitted with regression equations that can be used in quadratic programming. This approach to the formulation of maximum profit feeds is not presently used with fish, but it has as much merit for fish as it does with poultry.

PRACTICAL FEED INGREDIENTS

Ingredients used in practical fish feeds can be classed as protein (amino acid) sources; energy sources; essential lipid sources; vitamin supplements; mineral supplements; and special ingredients to enhance growth, pigmentation, or sexual development in the fish, or physical properties, palatability, or preservation of the feed. Some feed ingredients in common use worldwide and factors influencing their use in fish feeds are discussed in the following. Nutrient composition of these and other feedstuffs is presented in Appendix A.

Fish Meal

Fish meal made from good quality whole fish that is properly processed is the highest quality protein source commonly available to fish feed manufacturers. It is a rich source of energy and minerals, is highly digestible, and is highly palatable for most fishes. Fish meal made from whole fish contains 60% to 80% protein which is 80% to 95% digestible to fishes. It is high in available lysine and methionine, the two amino acids most deficient in plant feedstuffs. Marine fish meal contains 1% to 2.5% n-3 fatty acids, which are essential in the diet of many fishes. Fish meal made from fish parts, such as waste from fish processing and canning plants, is lower in percentage and quality of protein. It is also high in ash and should be used prudently in fish diets to prevent mineral imbalances which could cause inavailability of certain minerals.

Fish meal appears to be beneficial in fish diets in ways other than helping to meet the fish's minimum nutrient requirements. Research studies have shown that supplementation of either the lipid or the nonlipid fraction of fish meal to a nutritionally balanced channel catfish diet increased growth.

Because of its high cost, fish meal is generally used sparingly in commercial fish feeds, although some feeds, such as those fed to eels and pen-reared salmon, contain high levels of fish meal. Other animal protein sources, such as good quality blood meal, meat meal, or poultry byproduct meal, have been reported to replace some of the fish meal in experimental fish feeds.

Soybean Meal

Soybean protein has one of the best amino acid profiles of all protein-rich plant feedstuffs for meeting the essential amino acid requirements of fish. Data in Table 5.3 compare the available (digestible) essential amino acid contents of proteins from soybean, peanut, cottonseed, and menhaden fish meals, and the adequacy of each protein source for meeting the essential amino acid requirements of channel catfish and Japanese eel. Based on NRC (1983) requirements, soybean protein is not deficient in any essential amino acid for channel catfish. Because the reported methionine-plus-cystine requirement for eel is almost twice that for channel catfish, soybean protein is deficient in these amino acids and slightly deficient in threonine for eel. For eel, peanut and cottonseed proteins are deficient in all of the essential amino acids except arginine and phenylalanine plus tyrosine.

Some fish find soybean meal unpalatable. Chinook salmon will accept soybean meal better than coho, and older salmon accept it more readily than younger fish. Canadian research showed that soybean meal could comprise the major protein source in rainbow trout diets and produce satisfactory growth, but if the level of fish meal were reduced below approximately 18%, diet palatability was reduced. Channel catfish satisfactorily consume all-plant diets in which solvent-extracted soybean meal comprises 60% to 70% of the formula.

When commercially processed soybean meal replaces fish meal or other animal byproduct proteins in the diet, the losses in energy, minerals, and lipids should be considered. Dehulled soybean meal

Table 5.3. AVAILABLE ESSENTIAL AMINO ACID CONTENT OF SOYBEAN, PEANUT, COTTONSEED, AND MENHADEN FISHMEAL PROTEINS AND PERCENTAGE OF DIETARY REQUIREMENT FOR CHANNEL CATFISH AND JAPANESE EEL

| | Soybean | | | Peanut | | | Cottonseed | | | Fish meal | | |
| | Percentage of the protein | Percentage of the req. for: | | Percentage of the protein | Percentage of the req. for: | | Percentage of the protein | Percentage of the req. for: | | Percentage of the protein | Percentage of the req. for: | |
Amino acids		catfish	eel		catfish	eel		catfish	eel		catfish	eel
Arginine	7.25	168	161	9.10	212	202	9.17	213	203	5.59	130	124
Histidine	2.18	142	102	1.64	106	77[1]	1.87	121	88	2.01	130	95
Isoleucine	4.01	155	101	3.29	128	83	2.03	78	51	4.11	159	103
Leucine	6.35	181	181	5.17	148	97	3.90	111	74	6.53	186	123
Lysine	5.82	113	110	3.17	62	60	1.58	31	30	6.69	130	126
Methionine + cystine	2.52	108	50	2.03	87	40	1.86	80	37	3.15	135	70
Phenylalanine + tyrosine	7.08	141	121	6.89	138	118	6.01	120	103	6.30	126	108
Theonine	3.25	144	81	2.09	93	52	2.13	95	54	3.58	159	90
Tryptophan	1.18	219	111	0.88	162	83	0.86	160	81	0.91	168	86
Valine	4.09	138	103	3.51	118	88	2.94	99	73	4.59	155	115

Note: Underlined values are less than the requirement. Adapted from NRC (1983).

113

contains 25% less metabolizable energy (for rainbow trout), 86% less available phosphorus (for channel catfish), and 90% less n-3 fatty acids than anchovy fish meal.

Soybeans contain several anti-nutritional factors, namely a trypsin inhibitor. Heating during commercial oil extraction (approximately 105° C for 10–20 min.) destroys most of these factors. Several reports indicate that further heating improves the nutritional value of soybean meal for fish. However, altering conventional commercial soybean processing may not be economically practical.

Full-fat Soybeans

The use of full-fat soybean meal in fish feeds has received attention since research by the U.S. Fish and Wildlife Service showed that heating full-fat soybeans at high temperature ($\geq 177°$ C) improved the nutritional value for trout above that of commercial soybean meal. Canadian research, however, found that feeding a diet containing high levels ($\geq 80\%$) of full-fat soybean meal reduced weight gain for rainbow trout, presumably because it reduced consumption by the fish because of its higher energy and/or lower palatability. Trout fed full-fat soybean diets were significantly fatter. Channel catfish fed diets containing full-fat soybean meal gained the same amount of protein but much more fat than fish fed diets containing solvent extracted soybean meal. Full-fat soybean meal contains 18% fat compared with 0.5% fat for solvent-extracted soybean meal. The additional fat in the full-fat soybean meal is beneficial only if it improves the nutritional quality of the diet. Too much fat can cause an imbalance of protein and energy in the ration, which can reduce nutrient intake and also produce fatty fish. The additional fat in full-fat soybean meal would be more beneficial to coldwater fish, who do not utilize carbohydrates well for energy.

Cottonseed Meal

Cottonseed meal is an important source of protein in the United States and is available in many other parts of the world for animal feeds. Initially, formulated fish feeds in the United States contained significant amounts of cottonseed meal; however, after fish meal and other animal proteins decreased in content in commercial fish feeds, cottonseed meal was eliminated from many fish feed formulas because of its low available lysine content. Supplementing cottonseed meal with crystalline amino acids could possibly make it more competitive for use in fish feeds, because research has shown that the protein quality of peanut meal can be improved for channel catfish by supplementation with synthetic lysine.

Another liability of cottonseed meal is its content of free gossypol, which is moderately toxic to monogastric animals. Approximately 0.03% free gossypol will suppress growth rate in rainbow trout (Herman 1970). Levels of $\geq 20\%$ prepress solvent-extracted cottonseed meal in channel catfish diets cause gossypol toxicity, but lower levels do not. The amount of free gossypol in cottonseed meal depends upon processing (Lovell 1981). Generally, free gossypol contents of three types of cottonseed meal are as follows: direct solvent meal, 0.2% to 0.4%; screwpress meal, 0.02%; prepress solvent, 0.05%. The addition of 0.85 to 1.0 part ferrous sulfate to each part free gossypol in diets of swine and poultry has proven successful in blocking the toxic effects of gossypol.

Other Oilseed Meals

Meals from peanut (groundnut) and sunflower seed have been used in fish feeds in the United States. Compared with soybean meal, these are deficient in lysine and methionine. Peanut meal produced in the United States contains around 48% crude protein, is highly palatable, and contains no toxins or anti-nutritional factors. It also has better binding properties for pelleting than soybean meal. Sunflower meal is also palatable, but relatively fibrous. Canola meal, produced mainly in Canada, is comparable in protein quality to soybean meal but contains glucosinase enzymes, which hydrolyze glucosinolate to yield antithyroid products, thus making it a poor protein source for fish.

Grains and Byproducts

The primary nutritional contribution of grains is carbohydrate. Whole grains contain 62% to 72% starch, which is about 60% to 70% digestible by warmwater fish but less than 40% digestible by salmonids. Heating increases digestibility 10% to 15%. Starch in grains is a valuable binding agent in steam pelleted and extruded fish feeds.

Yellow corn is the most widely fed grain in North America. It is low in protein, and the protein is poor in amino acid balance. It contains approximately 20 mg/kg to 30 mg/kg of the yellow-pigmented xanthophylls (leutin and zeaxanthin), which impart undesirable yellow color in certain areas of the flesh of white fleshed fish. Approximately 11 mg of xanthophylls/kg feed can be used in catfish feeds without adverse effect on pigmentation of the flesh. Corn gluten meal contains 40% to 60% protein and is a good source of methionine; however, it is a highly concentrated source of xanthophylls (200–350 mg/kg). Corn byproducts from distilleries and breweries are relatively high in protein (26–28%) and fat (8–10%), but the protein is low in lysine.

Rice bran, which usually includes polishings, contains approximately 12% protein, 12% fiber, and 12% fat. It is often available at reasonable cost in developing countries where other ingredients may be prohibitively expensive. It is a good nutrient source, but does not pellet well because of the high fat and fiber contents.

Wheat is usually too valuable for human foods to feed to fish; however, small amounts of ground wheat are often used in fish pellets because of binding quality. Wheat gluten (protein) is an excellent binder. Wheat bran and middlings are commonly used in salmonid feeds because they contain more protein and less starch than whole grains.

Animal Byproducts

Meat and bone meal is a product of partially fat-extracted, dried slaughterhouse waste. It usually contains about 50% to 55% crude protein, with a significant amount of the protein coming from bone and other nonmuscle tissue. The quality of the protein is less than that of whole fish meal and varies considerably with the composition of the waste material. It is high in ash and therefore should not be used in unlimited quantity in fish feeds. It is a good source of energy, phosphorus, and trace minerals. Flash or spray-dried blood meal is highly digestible and rich in protein (80–86%). It is deficient in methionine but rich in lysine. It is not as good a mineral supplement as fish meal or meat and bone meal because of absence of bone. Blood meal is unpalatable to some species. Poultry byproduct meal, without feathers, is a good source of animal protein for fish feeds; however, this product is usually recycled back into poultry feeds. Feather meal is a highly concentrated source of protein (80%), but quality is poor and digestibility is low unless the feathers are thoroughly hydrolyzed during processing.

Crustacean Meals

Shrimp waste meal is a useful feed ingredient if the heads are included. The exoskeleton is primarily chitin and has limited nutritional value; thus the visceral organs in the head section are most valuable nutritionally. Crude protein analysis of shrimp waste meal must be corrected for the nitrogen in chitin, which accounts for about 10% to 15% of the total nitrogen. In addition to supplying protein, shrimp waste meal is a source of n-3 fatty acids, cholesterol (essential for crustaceans), and astaxanthin (for red pigment in salmonids). It is highly palatable and may serve as an attractant in feeds for fishes and crustaceans.

Krill (*Euphausia pacifica*), a small marine crustacean, is harvested from the sea and made into meal for fish, mammal, and poultry feeds. It usually contains less than 40% protein and is high in chitin. It is high in oil and is a rich source of n-3 fatty acids. It is a valuable source of the red carotenoid astaxanthin and is often used in fish diets for pigment enhancement in skin and flesh.

Fats and Oils

Fats and oils are used as energy sources, to provide essential fatty acids, and to coat the outside of pelleted feeds to reduce abrasiveness (resulting in less fines) and minimize dustiness. Fats from livestock and poultry processing (animal fats) are highly saturated, but are effective sources of energy for warm and coldwater fishes. Vegetable oils are generally higher priced than animal fats in the United States because of their value for human use, but are also good energy sources for fish. Marine fish oils are used in feeds for trout, salmon, and cultured marine fish because they are a source of essential (n-3) fatty acids. Marine fish oils contain on the average 20% to 25% of the long chain (over 20-carbon) n-3 fatty acids. Animal fats and vegetable oils do not contain fatty acids beyond 18 carbons and, except for soybean and linseed oils, most are low in n-3 fatty acids. Catfish oil, which is recycled back into catfish feeds, is similar to livestock and poultry fat in fatty acid composition. Fatty acid composition of fats and oils from various sources used in fish feeds is presented in Appendix A, Table A.4 (see page 253).

Fibrous Feedstuffs

Fish do not require fiber in the diet and, being monogastric animals, do not digest fibrous feeds well. It is unlikely that adding fiber to practical diets that contain 3% to 5% fiber will have any measurable benefit. In most cases, the concern is to avoid excessive levels of fiber in fish diets because high fiber levels reduce the binding quality of processed feeds, inhibit feed intake by fish, and increase fecal waste production.

Binding Agents

Steam pelleted fish feeds usually require the addition of special binding agents, whereas extruded feeds do not. Binding agents are especially valuable, if not essential, in crustacean feeds, which must remain water stable for several hours.

Table 5.4 presents binding agents used in commercial and laboratory pelleted fish feeds. Organic hydrocolloids, such as carboxymethylcellulose, guar gum, agar, and alginic acid, have been used successfully in laboratory diets but are expensive for commercial feeds. Commercial binding agents used in livestock and poultry feeds, such as hemicelluloses, molasses, and lignin sulfonates, are relatively inexpensive and are usually suitable for fish feeds that are consumed within 30 minutes after feeding; however, their effectiveness in feeds that must remain intact in water for several hours has been inconsistent. Inorganic binding agents, such as sodium and calcium bentonite or aluminum silicate, have been found to be less effective than organic binders for feeds fed in water.

Cooked (gelatinized) starches, used at $\geq 10\%$ of the diet formula, are good binding materials and are also sources of energy. They are highly hygroscopic, however, and will absorb water which causes the pellet to swell and disintegrate after a time in the water.

The polymethylocarbamide binder Basfin (BASF, West Germany) was used successfully in shrimp feeds. At levels of 0.5% to 0.8%, with good processing technology, it allowed pellets to remain intact in water for ≥ 6 hours. This compound is relatively nonhygroscopic and absorbs

Table 5.4. BINDING AGENTS USED IN STEAM PELLETED FISH FEEDS

Compound	Amount used %	Comment
Carboxymethylcellulose	0.5–2.0	Good, expensive
Alginates	0.8–3.0	Good in moist feeds, must combine with di-or polyvalent cation
Polymethylocarbamide	0.5–0.8	Very good, not FDA approved for use in United States, unpalatable to some fish
Guar gum	1–2	Good, expensive
Hemicellulose	2–3	Fair, moderate cost
Lignin sulfonate	2–4	Good, moderate cost
Na and Ca bentonite	2–3	Inferior to organic binders
Molasses	2–3	Fair, has nutritional value
Whey	1–3	Fair, has nutritional value
Gelatinized starches (corn, potato, sorghum, rice, cassava)	10–20	Good, large amount required, has nutritional value
Wheat glutin	2–4	Good, expensive

little water. It cross-links with carbohydrates (ether linkages) and proteins (N,N'-methylene bonds). However, dietary levels over 0.5% affected palatability in some fishes and it was removed from the market. Other compounds with similar binding properties are being researched.

PROCESSING FISH FEEDS

Pelleting and Extruding

Steam pelleting, through compression, produces a dense pellet that sinks rapidly in water. Extrusion is a process through which the feed material is moistened, precooked, expanded, extruded, and dried, producing low density feed particles that float in water. Steam pelleting is significantly less expensive than extrusion; however, extruded feeds are very popular with catfish farmers in the United States because they allow the feeder to observe the fish feed (see Figure 5.1).

Fig. 5.1.   Catfish feed mill in Mississippi which can process up to 1,000 tons of extruded (floating) feed daily. The feed is hauled in bulk directly to fish farms nearby.

Steam pelleting involves the use of moisture, heat, and pressure to agglomerate ingredients into larger homogeneous particles. Steam added to the ground feed mixture (mash) during pelleting partially gelatinizes starch, which aids in binding the ingredients. Generally, an amount of steam is added to the feed mixture to increase moisture content to approximately 15% to 18% and temperature to 70° C to 85° C before it is pressed through the pellet die. The moisture in the pellets is reduced by forcing nonheated air over the surface of the hot pellets immediately after they leave the pelleting apparatus (see Figure 5.2).

Fig. 5.2. Pellet Mill: (1) transmission; (2) feed conditioner (mixes steam); (3) feed distributor (to the pelleting chamber); (4) pelleting chamber; (5) adjustable feed plow; (6) pellet die ring; (7) die holding bolt; (8) die stiffener ring; (9) shaft; and (10) roll assembly (presses feed through holes in die ring).
(Courtesy of Sprout Waldron Co.)

Steam pelleted feeds must be firmly bonded for satisfactory water stability and to prevent disintegration if the pellets are crumbled into smaller particle sizes for small fish. Processing conditions and ingredient composition are both important to produce firmly bonded pellets. Fine grinding of all ingredients through a $\leq$ 2-mm screen or with an attrition mill prior to pelleting improves pellet stability. Starch, from grains, is important for good pellet binding. Fiber and fat are antagonistic to firm bonding in pellets. Thus, supplemental fats should not be added to the feed until after pelleting, and highly fibrous feedstuffs, such as wheat bran, should not be used in large quantity.

Special ingredients or binders are usually used in steam pelleted feeds for crustaceans, which must remain intact in the water for several hours. Precooked (pregelatinized) starches from corn, sorghum, potato, palm nut, and tapioca have been used at levels of 10% to 20% of the formula to increase bond strength of the pellets. Wheat gluten and ground wheat have good binding properties. Special organic hydrocolloids, such as those presented in Table 5.4, (page 118) are used in quantities of 0.5% to 3% of the ingredient formula.

Extrusion requires more elaborate equipment and more energy for processing than pelleting (see Figure 5.3). Higher levels of moisture, heat, and pressure are used. Usually, the mixture of finely ground ingredients is conditioned with steam or water into a "mash" which may or may not be precooked before entering the extruder. The mash, which contains around 25% moisture, is compacted and heated to 104° C to 148° C under pressure in the barrel of the extruder. As the material is squeezed through die holes at the end of the barrel, and external pressure decreases, part of the water in the superheated dough immediately vaporizes and causes expansion of the feed particles. The extruded particles contain more water than steam pelleted particles and require external heat for drying. Thus, after extrusion, the particles must pass through a drying tunnel to reduce moisture to a safe storage level. Heat-sensitive vitamins, especially ascorbic acid, are added in excess prior to processing or applied to the surface after processing. Extruded feeds are more firmly bound due to the almost complete gelatinization of the starch, and this results in fewer fines and longer water stability than pelleted feeds.

Meal and Crumbles

Fine particle or meal type feeds for small fish are usually prepared by pelleting the ingredient mixture and subsequently reducing the size of the pellets by crumbling or grinding. Initial grinding of the ingredi-

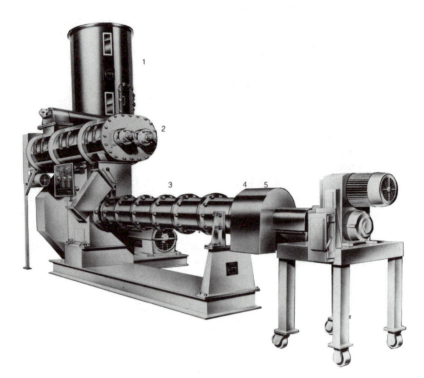

Fig. 5.3. Cooker-extruder for processing expanded (floating) fish feeds: (1) Bin-discharger; (2) preconditioner where steam and/or water are added; (3) extruder barrel where temperature is quickly elevated through friction, compression, and sometimes steam; (4) die for shaping particles; (5) knife for cutting off extruded particles.
(Courtesy of Wenger Manufacturing)

ents prior to pelleting to a very small particle size (≤ 0.5 mm) is important. Firm bonding of the pellet is necessary so that during regrinding, the pellet is reduced into homogeneous small particles and not disintegrated. The particles from the reground pellets are separated into various sizes by screening. If supplemental fat is included in the diet, it should be added after pelleting or after screening. Some meal type diets are not pelleted. The disadvantage of this procedure is that the water-soluble nutrients, such as some of the vitamins, dissolve very quickly after the feed is put into the water. Loss of water-soluble nutrients due to leaching is high with small particle diets, whether

Fig. 5.4. Various forms of fish feeds: (1) Extruded (floating); (2) steam pelleted; (3) large crumbles (from pellets); (4) small crumbles (from pellets); (5) extruded nonfloating (for shrimp); (6) flaked (drum dried); (7) fine meal (<0.5 mm); (8) coarse meal (0.5–1.0 mm).

prepelleted or not, and heavy fortification of water-soluble vitamins (two or three times the minimum requirements) is recommended. Fat-soluble vitamins should be added in a fat or oil carrier. The addition of fat to the surface of meal and crumble particles improves water stability, helps them to float, and reduces leaching of water-soluble nutrients.

Feeds for Larval Fishes

Many fishes begin feeding within a few days after hatching, before the digestive tract is fully developed. They have difficulty assimilating dry, prepared diets and are usually fed live foods such as unicellular algae, rotifers, and brine shrimp nauplii. Complete replacement of live foods with prepared diets for larval fishes has not been highly successful, but several feeds have been developed which partially replace the live foods.

Desired qualities of a larval diet are that the diet is nutritionally complete, palatable, and has desired physical properties. The particles should be 100 μm to 500 μm, have good water stability, and be relatively buoyant so they will be suspended in the water column for consumption by the fish.

Several methods are used to make larval feeds: pelleting with subsequent crumbling to reduce particle size; flaking and regrinding to small particles; and microencapsulation. Fine grinding of ingredients is very important in all processes. An attrition mill is often used to grind ingredients to sizes of $\leq$ 200 μm. Larval feeds made by pelleting are high density and sink relatively fast when fed. The pellets are crumbled or reground into smaller particles which are screened into various size groups ranging from 100 μm to 500 μm. The flaking process involves processing on a drum dryer, as described later, grinding to small particles, and sizing the particles. These particles usually have less density than the pelleted particles and remain in the water column better.

Microencapsulation involves coating a small particle or beadlet of diet with a thin layer of a compound that will reduce dissolving, leaching, or in some cases bacterial degradation, until the material is consumed by the fish or removed from the rearing container. There are several published and patented processes for microencapsulation, and these vary with the encapsulation material and the substrate being coated. Nylon (N-N bonds) cross-linked proteins, calcium alginate, and oils have been used as encapsulation materials. The materials should be water insoluble but digestible by enzymes in the digestive tract of the larval fish. For more detail on microencapsulation processes, the reader may refer to Jones and Gabbott (1976).

Moist Diets

Moist feeds are prepared by adding moisture and a hydrocolloidal binding agent (e.g., carboxymethylcellulose, gelatinized starch, gelatin) or fresh tissue (e.g., liver, blood, ground fish or fish processing waste) with the dry ingredients and extruding to form moist pellets. Advantages of moist feeds are that many fish species find soft diets more palatable than dry diets, a pelleting machine is not needed (a food grinder will suffice), and heating and drying are avoided. Disadvantages are that wet feeds are susceptible to microorganism or oxidation spoilage unless fed immediately or frozen. Fish or fish parts going into moist feeds should be heat processed to destroy possible pathogens and thiaminase, the thiamine-destroying enzyme found in most fish tissues. The Oregon Moist Pellet (discussed in chapter 9 under Practical Feeding—Trout and Salmon), which was developed for salmon smolts, is an example of a commercial moist feed. It is stored frozen.

Some moist diets do not require frozen storage. They contain humectants like propylene glycol and sodium chloride which lower water activity below that which will allow bacteria growth. They also contain fungistats like propionic acid or sorbic acid, which retard mold growth. These diets must be packaged in hermetically sealed containers and stored at low temperatures for best storage life. The moisture enhances loss of vitamin C.

Eel feeds are processed and stored dry but moistened just before feeding. These feeds contain 10% to 20% pregelatinized starch, which serves as a binding agent. About 5% to 6% fish oil and 50% to 100% water is added to the dry mix. The moist mixture can be fed in large balls or extruded through a food grinder into smaller particles.

Flaked Diets

Aquarium fishes require diets that are not only nutritious and palatable, but that also float or sink slowly and will not disintegrate quickly in water. Flaked feeds processed on rotary drum dryers have met these criteria. The ingredients are ground to extremely fine particle size (approximately 0.1 mm) with equipment such as an attrition mill and blended with water to form a slurry that is spread over the surface of a heated rotating cylinder (drum) to dry into a thin sheet. The dried sheet is continuously scraped off the rotating drum and crumbled into flakes.

The formula must contain ingredients with good hydrocolloidal properties as well as tensile strength. Boonyaratpalin and Lovell (1977) have discussed ingredient and processing factors important in processing flaked diets. They found that chitin, from shrimp shells, was important in imparting the desired physical properties to the flakes. They also found that much of the vitamin C is lost.

Compounds for external pigment enhancement are often added to aquarium feeds. Astaxanthin (in crustacean meal) and canthaxanthin (a synthetic) impart pink-red color, while xanthophylls from plant pigments can supply yellow-orange pigmentation.

Crustacean Diets

Crustaceans are relatively slow feeders and require diets that will remain stable in water for a much longer time than those for most finfish. Ingredients with good binding properties and with special binding agents should be used in stream pelleted crustacean feeds. Extrusion processing is also a valuable tool in making water-stable crustacean feeds, provided the amount of expansion of the feeds is restricted so they will not float; however, extrusion is expensive and may not be available in many areas where crustaceans are cultured. Good quality shrimp feeds can be processed with properly designed pelleting equipment and by using good processing technology. Some pelleting machines can be fitted with special die rings that allow the feed mixture to receive more heat during compression; this causes more gelatinization of starches. The diameter of the pellets should be relatively small (3–5 mm); this will allow for more heated surface and for optimum consumption by the shrimp.

Pond Feeds for Bait and Ornamental Fish

Meal type feeds are often used in ponds for feeding bait fishes, such as golden shiners and fathead minnows, and for feeding ornamental fishes. Sometimes the meal type feeds are fed until the fish are harvested, which is usually at a maximum size of 10 cm to 12 cm. Greatest feed efficiency is obtained, however, when the fish are changed to small-particle (2–4 mm), crumbled or extruded feeds. The extruded feed particles are carefully processed to insure sufficient expansion to allow them to float but to keep particle size small.

Most bait and ornamental fishes can obtain a large part of their nourishment from plankton and benthic organisms in the ponds; consequently, good pond preparation is important. When fish consume a large amount of natural food in a pond, the micronutrient composition of the supplemental diet is not critical.

REFERENCES

ARRAES, R. A. 1983. Alternative evaluations of economically optimal rations for broilers. Ph.D. diss., University of Georgia, Athens, GA.

BOONYARATPALIN, M., and R. T. LOVELL. 1977. Diet preparation for aquarium fishes. *Aquaculture* 12: 53–62.

HERMAN, R. L. 1970. Effects of gossypol on rainbow trout *Salmo gairdneri. J. Fish. Biol.* 2: 293–303.

JONES, D. A., and P. A. GABBOTT. 1976. Prospects for the use of microcapsules as food particles for marine particulate feeders. In *Proceedings of the Second International Symposium on Microencapsulation,* ed. J. D. Nixon. New York: Marcel Decker, Inc.

LOVELL, R. T. 1981. Cottonseed meal in fish feeds. *Feedstuffs* 53: 28–29.

National Research Council. 1983. Nutrient requirements of warmwater fishes and shellfishes. National Academy of Sciences. Washington, DC: National Academy Press.

SANTIAGO, C. 1985. Amino acid requirements of *Tilapia nilotica*. Ph.D. diss., Auburn University, Auburn, AL.

6

Fish Feeding Experiments

Many fish feeding experiments must be conducted in a controlled environment to prevent interaction of environmental effects, such as natural food organisms, temperature, and water quality, with the nutrient variable being studied in the experiment. For example, nutrient requirements cannot be determined unless nutrient consumption by fish is precisely known. Such studies are conducted in aquaria or tanks, indoors, with controlled temperature and water quality. Some studies, however, which involve evaluation of practical feed formulations or feeding regimes, should be conducted under conditions as similar as possible to the conditions where the results will be applied while at the same time allowing for accurate collection and analysis of data. These studies are conducted in experimental ponds, raceways, pens, and similar enclosures of water.

CONTROLLED ENVIRONMENT STUDIES

The requirement or effect of a specific nutrient or compound in the diet must be determined in a controlled environment. The overriding consideration in any feeding experiment is to keep all controllable variables equal among treatments except the one being tested. In formulating experimental diets, the physical, palatability, and nutritional properties of each should be equal except for the nutrient or variable that is being tested. Characteristics of controlled environment feeding studies that will yield reliable and useful information are presented in the following.

Rearing Facilities

Size of rearing containers should be large enough to accommodate the originally stocked population after a 500% to 1,000% weight increase. Containers should be supplied with a continuous flow of good quality water with temperature regulated as desired or to that accepted as standard for the fish species (see Figure 6.1). If surface water is used, it should be filtered to remove all natural sources of nutrients. A constant diurnal light/dark cycle of approximately 14:10 should be maintained for the experimental fish.

Test Fish

Full siblings of a fast-growing genetic strain are desirable. Small fish respond faster than large fish to nutritional variables. Also, small fish are more sensitive to diet differences; if small fish are unaffected, it is a safe assumption that larger fish will not be. Higher numbers of small fish per rearing container can be used. However, where nutrient or energy requirements will change with size, different fish size groups should be evaluated. The minimum number of fish per rearing unit (tank or aquarium) should be high enough to negate effects of unequal sex ratio (males growing faster) or to prevent hierarchical feeding patterns. For statistical purposes, each treatment should be replicated in a minimum of three rearing containers.

Previous history of the experimental fish is important. Compensatory growth and tissue stores of nutrients can have significant effects on response to test diets. Shell (1963) showed that there is growth compensation in channel catfish. Two groups of fish were fed experi-

Fig. 6.1. Experiments to evaluate the nutrition of the fish must be conducted in a controlled environment like the laboratory above, where the aquariums can have a continuous flow of temperature-controlled water that is free of nutrient sources and aeration to keep dissolved oxygen constant.

mental diets, one that caused a significantly lower growth rate than the other. In a subsequent experiment, when both groups were placed on a different, high quality diet, the group that grew slowly in the first feeding trial grew faster than the other group in the second trial.

In cases where the metabolic requirement for a nutrient is extremely low and the nutrient has a long residual time in fish tissues, such as with essential fatty acids, body stores of the nutrient may preclude use of the fish in the experiment without depleting the fish prior to the experiment. In vitamin B_{12} studies with chickens, only second generation birds from parents fed B_{12}-free diets could be used.

Test Diets

Experimental diets for evaluating nutrient requirements should be prepared from highly purified ingredients to allow maximum control over the nutrient being tested. Preferably, all diets in an experiment should be:

1. alike in all respects except the variable being tested;
2. palatable;
3. feedable (water stable, optimum particle size and texture);
4. nutritionally complete (except for the nutrient being tested); and
5. as far as possible, made from highly purified ingredients.

Casein and gelatin (4 : 1) are a good protein combination for purified diets; vitamin-free casein is available for use in vitamin experiments. Blood fibrin is a desirable protein for mineral studies. Egg protein is good for use in protein or amino acid experiments. All of these protein sources are available in highly purified forms.

Dextrin is traditionally used as a carbohydrate source. Cooked starch is satisfactory for warmwater fishes, but starch is not utilized as well as dextrin by coldwater species. Some of the lipid should be from fish oil to provide n-3 fatty acids; the remainder, which is primarily an energy source, may come from vegetable or animal fats. Purified cellulose is used as a nonnutritive filler.

Whether processed into moist or dry pellets, the feeds must contain a binding agent that will hold the particles together for a reasonable time in water. Gelatin, agar, alginic acid, or carboxymethylcellulose are used at levels of 2% to 5%. Steps for preparing semipurified moist diets are as follows:

1. Mix the dry ingredients well before adding oil, then add oil and mix.
2. Add approximately 350 ml water/kg diet mixture and stir; the moist mixture should have a stiff, plastic consistency when compressed. If it does not stick together, add more water, but too much water will cause sticking in the food grinder.
3. Extrude through a food grinder with proper diameter holes in the grinder plate (Figure 6.2).
4. Break extrusions into short lengths by hand or with a sharp implement, package, and freeze quickly.

Fig. 6.2. A laboratory fish diet made from highly purified ingredients and mixed with 25% to 40% water is being extruded through a food grinder. The diet can be dried or frozen and fed moist.

Moist diets should be stored frozen until 1 or 2 days prior to feeding and should be kept refrigerated during this period to minimize loss of nutrients sensitive to degradation.

Model semipurified diets (made from highly purified ingredients) for warmwater fishes, salmonids, and crustaceans are presented in Tables 6.1 through 6.3. A standard mineral mixture for semipurified fish diets has not been developed. That presented in Table 6.4 is proposed to provide the minimum requirements for essential elements for fish. The vitamin mixture used in the semipurified diet for warmwater fishes presented in Table 6.1 has been used in research diets with channel catfish, tilapia, carps, largemouth bass, and ornamental fishes.

In experiments where indigenous nutrients or other components of the diet ingredients will not confound the experiment, practical feedstuffs may be used in the basal diet. If the source of the nutrient being tested is a crude feedstuff, its availability to the fish must be known. If a commercial feed is used as the basal diet, it should be finely ground, resupplemented with vitamins, and reformed into optimum size particles for feeding.

To estimate the amount of each diet required for the experiment, estimate the amount of weight the fish will gain during the experiment and assume 1.5 g of dry diet is required for 1 g weight gain. Thus, if weight is expected to increase 500%, amount of diet needed would equal (1.5)(5)(initial weight of total fish per treatment).

Table 6.1. EXAMPLE OF A SEMIPURIFIED TEST DIET FOR WARMWATER FINFISH

Ingredient	Percentage of diet
Casein, vitamin-deficient	32
Gelatin	8
Dextrin	28
Cellulose	19
Marine fish oil and/or animal fat[1]	6
Carboxymethylcellulose	2
Mineral mixture[2]	4
Vitamin mixture[3]	1

[1] Marine fish oil, animal fat usually in 1:1 ratio. Stripped lard is the sole lipid used in studies involving fat-soluble vitamins.
[2] Use fish mineral mix in Table 6.4.
[3] The mixture is diluted in cellulose and provides the following vitamin activities in mg or IU/kg of diet: vitamin A, 3,000 IU; vitamin D_3, 1,500 IU; vitamin E, 50 IU; menadione, 10; choline, 2,000; niacin, 50; riboflavin, 20; pyridoxine, 10; thiamine, 10; pantothenic acid, 40; folic acid, 5; vitamin B_{12}, 0.02; biotin, 1; inositol, 400; vitamin C, 200.
Source: from Lovell, Miyazaki, and Rebegnator (1984). Diet provides 36% crude protein, 3.0 kcal digestible energy/g.

Table 6.2. EXAMPLE OF A SEMIPURIFIED TEST DIET FOR COLDWATER FINFISH

Ingredient	Percentage of diet
Casein, vitamin-deficient	40.0
Gelatin	10.0
Oil, marine fish or soybean[1]	10.0
Sucrose	10.0
Dextrin	10.0
Cellulose	12.0
Minerals[2]	4.0
Vitamins[3]	0.9

[1] Marine fish oil is the preferred source of n-3 fatty acids, although soybean oil is satisfactory for rainbow trout.
[2] Use fish mineral mix in Table 6.4.
[3] Provides the following in IU or mg/kg of diet: vitamin A, 10,000 IU; vitamin D_3, 4,000 IU; vitamin E, 75 IU; vitamin K, 22; thiamine, 40; riboflavin, 30; D-calcium pantothenate, 150; niacin, 300; pyridoxine, 20; folic acid, 15; vitamin B_{12}, 0.3; inositol, 500; biotin, 1; vitamin C, 200; choline, 3,000.
Source: from Poston (1976) and NRC (1973).

Table 6.3. SEMIPURIFIED CRUSTACEAN REFERENCE DIETS

Ingredient	Casein base[1] (%)	Crab protein base[2] (%)
Vitamin-free casein	31	—
Purified crab protein	—	40
Corn starch	24	15
Cellulose	15	18
Wheat gluten	5	5
Dextrin	—	5
Corn oil	2	3
Cod liver oil	4	6
Soy lecithin	10	—
Spray-dried egg white	4	—
Choline chloride	—	1.0
Cholesterol	0.5	1.0
Mineral premix[3]	3	4
Vitamin premix[4]	2	2

[1] Developed at the Bodega Marine Laboratory, Bodega Bay, CA, USA.
[2] Developed at the Halifax Fisheries Research Laboratory, Halifax, Nova Scotia, Canada.
[3] Bernhart-Tomarelli modified salt mix (U.S. Biochemicals, Inc., Cleveland, OH, USA).
[4] Containing all of the vitamins listed in Table 2.6, chapter 2 (see page 33), in amounts recommended for salmonids.
Source: Crustacean Nutrition Newsletter, vol. 3, no. 1, June 15, 1986, Department of Fisheries and Oceans, Halifax, Nova Scotia, Canada.

Table 6.4. MINERAL PREMIX FOR SEMIPURIFIED DIETS FOR FISHES. TO BE USED AT THE RATE OF 4% OF THE DRY DIET

Ingredient	Percentage of premix
Aluminum potassium sulfate	0.159
Calcium carbonate	18.101
Calcium diphosphate	44.601
Cupric sulfate·5H$_2$O	0.075
Cobaltous chloride	0.070
Ferric citrate·5H$_2$O	1.338
Magnesium sulfate	5.216
Manganese sulfate·H$_2$O	0.070
Potassium chloride	16.553
Potassium iodide	0.014
Zinc carbonate	0.192
Sodium diphosphate	13.605
Sodium selenite	0.006

Note: Mineral premix provides the following amounts in mg/kg of the dry diet: aluminum, 7; calcium, 8,140; chloride, 6,008; cobalt, 12; copper, 8; iron, 104; iodine, 4; magnesium, 421; manganese, 10; phosphorus, 5,250; potassium, 3,474; sodium, 1,932; selenium, 1; zinc, 40.

Management

The fish should be conditioned to the rearing environment for 1 to 2 weeks prior to beginning the experiment. During this time they can be given chemical baths for external pathogens and acclimated to experimental diets. Sometimes it is necessary to feed all of the experimental diets for a preliminary period to be sure there is no difference in diet acceptability.

During the experiment, the fish should be fed for maximum growth rate. This enhances sensitivity to diet differences. Also, fish response may be different if the fish are underfed. Thus, satiation feeding is essential. Small fish may be fed several times daily; food-size fish should be fed once daily or according to practice. Length of the feeding trial will depend upon the time required to get statistical differences among the test diets. This will be influenced by the sensitivity of the fish to the limiting nutrient tested, fish species, size of fish, and other factors that affect growth rate.

If disease or other environmental problems not related to the diet occur in any of the rearing containers, feeding in all containers should be discontinued until the affected fish return to normal feeding; otherwise unequal feeding activity among sick and healthy fish may confound treatment affects.

Fish should be sampled at the beginning of the experiment for analysis for initial condition or composition of the fish. Biweekly weighings will allow comparison of weight gain over the course of the experiment; this will also be useful in feeding the fish.

Measuring Responses

Growth is usually the most important criterion for measuring fish response to experimental diets. In research studies such as determining nutrient requirements, growth is a sensitive and practical indicator of the adequacy of the diet for a particular nutrient or energy level. True growth in animals involves an increase in structural tissues such as muscle, bone, and the organs. This should be distinguished from an increase in fat deposition in the reserve tissue. Thus, growth can be characterized reliably by an increase in protein because muscle, organs, and demineralized bone tissue contain primarily protein. It is often assumed that weight gain is synonymous with growth or that composition of gain (protein and fat gain) is the same among fish fed different diets. This, however, is not a safe assumption because diet composition can influence how much of the absorbed nutrients go to protein and how much go to fat.

Weight gain is a reliable indicator for growth as long as the experimental variable is not expected to affect composition of gain in the fish. But if the variable being tested does affect composition of gain, erroneous conclusions can be drawn from weight gain results. An example is a feeding study that was conducted at Auburn University to determine energy and protein requirements of channel catfish. The fish were fed diets containing five levels of digestible energy (2.2–4.6 kcal/g) at each of two protein percentages (27 and 37%). They were allowed to eat as much as they would consume. As shown in Figure 6.3, weight gain increased linearly as digestible energy in the diet increased at both protein levels. However, protein gain increased with dietary energy only at the higher protein level. Thus, weight gain would indicate that additional energy improved both diets for fish growth; but measurement of composition of gain showed that the fish fed the lower protein diet used the additional energy for fat deposition instead of protein gain.

Unless there is reason to justify not measuring composition of gain, fish from both aquarium and production type experiments should be chemically analyzed at the beginning and end of the feeding trials for fat, protein, and moisture. If there is a difference in final body composition of fish fed the different diets, growth should be reported as

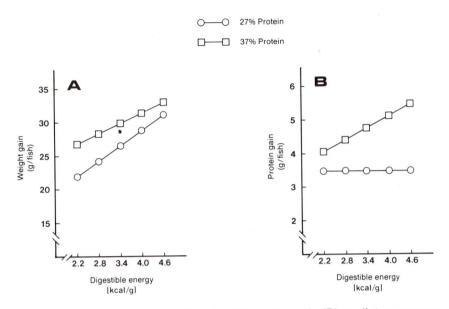

Fig. 6.3.   Regressions of gains in weight (A) and protein (B) on dietary energy concentration at two protein levels in channel catfish.
(From Mangalik 1986)

protein gain. Fortunately, fish are small animals, in contrast to warmblooded food animals, and whole body analysis is possible.

Presence of clinical signs may be indicators of treatment effects in nutrition experiments. Lesions, hemorrhage, abnormalities in pigmentation, gill hyperplasia, cataracts, and skeletal and cartilage deformities are things to be observed. Subclinical signs or measurements, such as enzyme activity, tissue or cell damage (histopathology), tissue levels of the test nutrient or its metabolites, and a number of others, are often more sensitive than growth response and should complement growth in evaluating the experimental effects. Specific subclinical indicators for various nutrient deficiencies are discussed in chapter 2.

Quantification of Nutrient Requirements

Determination of the quantity of a nutrient required in the diet by the fish generally involves feeding a series of diets containing various levels of the tested nutrient in a controlled environment and finding the lowest level that will elicit a desired rate of response from the fish.

Statistical handling of the diet response data is as important as any other part of the research. Traditionally, the lowest level of the test nutrient fed that produces the maximum response in a series of diets containing different levels of the nutrient is considered to be the required level. Regression analysis by the broken line method of Robbins, Norton and Baker (1979), is often used. The point of intercept of the lines representing the linear and horizontal parts of the growth curve is derived mathematically. However, fish response, such as growth, will generally show sharp initial increase as nutrient level in the diet increases, then a slower rate of increase with gradual leveling off, and subsequently (but not always) a decrease at the higher nutrient levels. Choosing the point on the growth response curve that represents the minimum level of nutrient for the desired rate of growth is critical because the nutrient level for maximum growth may be notably higher than that for an acceptable rate of growth. For nutrients where the recommended dietary level has greater physiological than economical importance, as for most micronutrients, maximum growth is a suitable response point. But for nutrients that are of significant economic importance, such as amino acids, a response less than maximum may result in significant reduction in feed cost without significant loss in growth rate.

Quadratic regression analysis of growth response data can yield nutrient requirements for maximum and less than maximum rates of growth. The growth curve in Figure 6.4 represents response of Nile tilapia to dietary levels of the amino acid arginine ranging from below to above the requirement for optimum growth (Santiago 1985). X_{max} on the quadratic regression curve represents the level of the amino acid associated with maximum growth (Y_{max}); X_1 represents the quantity of arginine associated with a rate of growth lower but within the 95% confidence limit of Y_{max}. The level of arginine at X_{max} is 4.4% of the protein, while the level at X_1 is 3.5%, 20% less. Thus, 20% less arginine can be fed with probably a significant reduction in feed cost with only a 5% likelihood of reducing growth rate.

PRACTICAL ENVIRONMENT STUDIES

Research to evaluate practical diets or feeding regimes should involve fish fed in an environment similar to commercial practice, but which is amenable to sound statistical design.

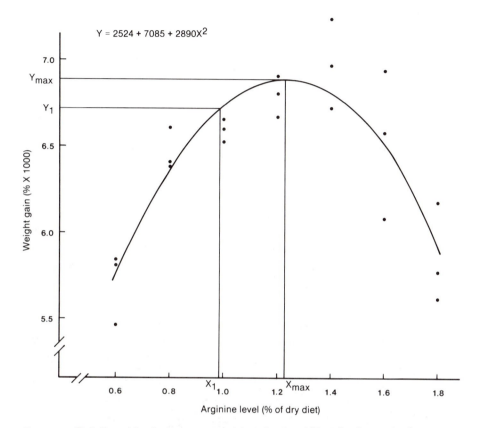

Fig. 6.4. Relationship between weight gain by Nile tilapia and dietary arginine level as described by quadratic regression, which allows derivation of X_{max}, which is the requirement for maximum growth (Y_{max}), and X_1 which is the requirement associated with a rate of growth below maximum (Y_1) but within the 95% confidence interval.
(From Santiago 1985)

Pond Experiments

Pond feeds and feeding practices should be evaluated in experimental ponds. The results of pond experiments represent the combined and inseparable effects of nutrients from the pond and the test feed. Some fish species consume relatively small amounts of pond organisms, while others consume significant amounts. Channel catfish will gain approximately 250 kg/hectare in ponds without feeding, while tilapia can gain 1,000 kg/hectare to 2,000 kg/hectare from natural pond food alone (Lovell, Shell, and Smitherman 1978). Other confounding effects of pond environments for feeding experiments are changes in temperature and water quality. Water quality usually deteriorates as the feeding period progresses and this may influence feeding activity of fish or may interact with experimental diets or feeding regimes. However, these environmental effects are characteristic of pond cultures and will be present in practice. The researcher should be cautious about transposing results of pond experiments to other culture environments, such as raceways.

Earthen Ponds.   An experimental pond should represent a commercial pond. Minimum size should be 0.1 hectare. Concrete or plastic-lined ponds do not represent earthen culture ponds. A source of water should be available to replace evaporation loss or to flush the pond in the event of oxygen depletion, and to rapidly refill ponds when restocking. The ponds should be easily drained and harvested. See Figure 6.5, which shows a series of earthen ponds.

Fig. 6.5. Earthen ponds are used to evaluate pond feeds and feeding practices. This is an aerial view of a series of research ponds ranging from 0.04 to 0.4 hectares in size at Auburn University, Auburn, Alabama.

Number of Ponds.   Because of great variation among individual ponds managed similarly, a large number of replicates are needed per treatment to discern meaningful differences among treatments. Shell (1983) reported that the coefficient of variation (100 × standard deviation/mean) of experimental ponds treated alike was 10 to 15 for channel catfish ponds in Alabama and 8 to 16 for common carp ponds in Israel. Thus, small differences in treatment effects on fish production in ponds are not practical to measure without an excessively (and expensively) large number of ponds. With knowledge of the amount of experimental error associated with a set of experimental ponds, the number of replications required to demonstrate a desired percentage difference among treatments can be determined statistically. Ordinarily, a minimum of three replicate ponds per treatment are used, and because of costs, more than three or four ponds per treatment are seldom used. Thus, small differences should not be sought using earthen ponds.

Management.   Experimental ponds should be stocked and managed similarly to existing or potential commercial conditions while allowing for accurate collection of data. In most cases, the fish should be fed over a growing season. Because of changes in temperature and water quality, a short-term feeding trial over a limited segment of the normal growing period may not give results applicable to practical conditions.

Feeding.   Fish in pond experiments should be fed as much as they will consume. Feeding at restricted levels can prevent the full benefit of the superior diets from being realized. Another problem with restricted feeding is that most fish species exhibit hierarchial feeding order and when a limited amount of feed is fed, the more aggressive feeders will consume as much feed as they want while some fish in the pond will get little. Most commercial fish farmers feed as much as the fish will consume to obtain maximum growth rate. Another advantage of feeding as much as the fish will consume is that if the numbers of fish in the experimental ponds become unequal, due to predation, mortality, stocking discrepancy, or other reasons, satiation feeding still insures all fish in the pond an opportunity to eat as much as they want.

Feeding fish to satiation can be done conveniently if floating (extruded) feeds are used; or by following reliable feeding tables, such as those developed for channel catfish and rainbow trout, if sinking feeds are used. Floating feeds allow the researcher to see how much the fish are consuming and remove unconsumed feed. When sinking pellets are used, the fish should be sampled periodically to adjust the feed allowance according to the feed allowance tables. Feed allowances should be adjusted at least once weekly. Demand feeders are another means of allowing the fish to eat as much as they will consume. Overfeeding will also allow the fish to consume as much as they want, but feed efficiency cannot be measured.

Experimental feeds should be processed into pellets of such diameter that the young, initially stocked fish can consume them. Dry pellets can be processed with commercial steam pelleting or extrusion equipment. If experimental feeds are made by agreement with commercial mills, the researcher should personally oversee the mixing, processing, and packaging of the feeds. This eliminates uncertainty about possible error in preparing the feeds.

Evaluating Fish Response.   Periodic sampling of fish from pond feeding experiments is sometimes necessary for adjusting feed allowance or for associating time with fish response; however, with some species, sampling stresses the fish and makes them more sensitive to disease and causes them to stop eating.

Weight gain is usually considered the most important measurement of the productivity of experimental feeds. Here, it is assumed that the weight of the fish is a reliable indicator of "marketable product." Fish weight represents marketable product to the producer, but not necessarily to the processor. An example of misrepresentation of the value of the diet by weight gain measurement can be seen from data from a channel catfish pond feeding trial (Deru 1985). The fish were fed as much as they would consume of isocaloric diets (2.88 kcal DE/g) at two protein percentages (26 and 32%) from fingerlings to harvestable size. As shown in Table 6.5, weight gain was almost the same for two protein levels, but the fish fed the lower protein diet gained 24% more

Table 6.5. EFFECTS OF ISOCALORIC DIETS OF TWO PROTEIN LEVELS ON WEIGHT, FAT AND PROTEIN GAINS BY CHANNEL CATFISH IN PONDS

Diet		Weight gain (g)	Fat gain (g)	Dressing percentage
Protein (%)	Digestible energy (kcal/g)			
26	2.88	254	41	54.5
32	2.88	258	33	56.6

Source: Deru (1985)

fat and had a 3.9% lower dressing percentage. If this feeding trial were evaluated on the basis of weight gain alone, the lower protein feed would be recommended because of lower cost. But, if yield of marketable carcass and quality of the final food product were considered, it may not be recommended.

Weight gain, or protein gain if true growth rate is desired, can be presented in many ways: standing crop, net production, percentage gain, gain per day, or growth curves. Standing crop, which is the total weight of the fish in the pond, is meaningful to the commercial culturist in that this indicates the total weight of fish that can be harvested in terms of yield per hectare of the culture system. If there are unequal numbers of fish in the ponds, fish response should be reported as average gain per fish. Net production represents final weight minus initial weight and can be presented on a per pond (hectare) or per fish basis. Percentage gain gives an indication of how much the fish has increased in size in relation to its initial weight or the weight of a control treatment. It is usually of more interest to a researcher than a farmer.

Average daily gain is a commonly used measure of livestock responses in feeding trials. It has limited application to fish feeding experiments because daily gain is related to size of fish; large fish gain more grams per day than small ones. Thus, data must be collected under standardized conditions with regard to fish stocking size, length of feeding trial, temperature, and so on, and at present there are no standardized conditions for fish "feedlot" studies.

Other desirable measurements for pond feeding experiments are fish size variation within a pond or treatment (the number or percentage of fish in various size ranges), feed conversion ratio, and possible clinical and subclinical nutrient deficiency signs.

Feed conversion or efficiency (reciprocal of conversion) cannot be accurately determined if the amount of feed consumed cannot be accurately ascertained. Thus, unless the researcher has a measure or reliable estimate of the net amount of feed consumed, he or she should not bother calculating feed conversion coefficients.

Raceways and Pens

Raceways, tanks, pens, and cages, which simulate commercial culture conditions, may be used as experimental growing units to evaluate practical feeds or feeding regimes. These units represent highly artificial culture conditions, and environmental interaction with the test feeds is usually minimal. The experimental diets must be nutritionally complete. Variation in fish response among rearing units within treatments will be less than in earthen ponds, thus smaller differences among test feeds should be detectable in these units than in ponds using the same number of replicates.

REFERENCES

DERU, J. 1985. Composition of gain of channel catfish fed diets containing various levels of fat at two protein percentages. Master's thesis, Auburn University, Auburn, AL.

KANAZAWA, A., S. TESHIMA, and S. TAKIWA. 1977. Nutritional requirements of prawn. VII. Effects of dietary lipids on growth. *Bull. Jap. Soc. Sci. Fish.* 43: 849–856.

LOVELL, R. T., T. MIYAZAKI, and S. REBEGNATOR. 1984. Requirement for alpha-tocopherol by channel catfish fed diets low in polyunsaturated diets. *J. Nutr.* 114: 894–901.

LOVELL, R. T., E. W. SHELL, and R. O. SMITHERMAN. 1978. Progress and prospects in fish farming. In *New protein foods,* eds. A. M. Altschul and H. L. Wilke, pp. 261–292. New York: Academic Press.

National Research Council. 1973. Nutrient requirements of coldwater fish. Washington, D.C.: National Academy of Sciences.

POSTON, H. A. 1976. Optimum level of dietary biotin for growth, feed utilization, and swimming stamina of fingerling lake trout. *J. Fish. Res. Board Can.* 33: 1803–1806.

ROBBINS, K. R., H. W. NORTON, and D. H. BAKER. 1979. Estimation of nutrient requirements from growth data. *J. Nutr.* 109: 1710–1714.

SANTIAGO, C. R. 1985. Essential amino acid requirements of Nile tilapia. Doctoral diss., Auburn University, Auburn, AL.

SHELL, E. W. 1963. Effects of changed diets on growth of channel catfish. *Trans. Amer. Fish. Soc.* 92: 432–434.

SHELL, E. W. 1983. Fish farming research Auburn, AL: Alabama Agricultural Experiments Station.

7

Practical Feeding—
Channel Catfish

Culture of channel catfish (Figure 7.1) is the largest aquacultural industry in the United States. It grew from insignificance in the late 1960s to an annual production of near 150,000 tons in 1986 (USDA 1987). The freshwater catfish carries a stigma in the eyes of many consumers in the United States. However, energetic market development has increased the demand for this fish in many areas. The flesh is mostly white muscle, is free of intramuscular bones, and has a mild flavor.

Catfish farming began in the southeastern United States in ponds used for sport fishing. Because channel catfish have long been popular in that area, they were grown and processed for retail food markets. Earthen ponds were stocked with 2,500 to 5,000 fingerlings per hectare in early spring; the fish were fed pelleted, concentrated feeds and were harvested the following fall. Early yields were 1,000 kg/hectare to 2,000 kg/hectare. By increasing stocking densities, improving nutrition, compensating for water quality problems, and using multiple harvesting techniques (harvesting the large fish several times per season and simultaneously restocking with small fish, without draining the pond), yields have been increased to 4,000 kg/hectare to 5,000 kg/hectare per year. On the Mississippi River flood plain, where large ponds can be built economically, catfish farming has become a major enterprise. Many farms are several hundred hectares in size, with individual ponds of 5 hectares to 10 hectares.

The channel catfish has many desirable traits for intensive culture. It can be spawned in captivity and can be cultured in ponds or in densely stocked cages or raceways. It grows rapidly; a 10-g fingerling reaches a harvest size of 0.5 kg in 6 months if the water temperature remains above 23° C. It accepts a variety of supplemental feeds and is relatively disease-free when environmental stresses are minimized. It tolerates daily and seasonal variations in pond water quality and temperature (from near freezing to 34° C). Although it makes efficient

Fig. 7.1. Channel catfish, approximately 0.5 kg in size.

weight gains from processed feeds, it does not make commercially economical gains in ponds without supplemental feeding.

The major problem in the culture of catfish in ponds is insufficient dissolved oxygen. Phytoplankton cannot produce enough oxygen during prolonged cloudy periods to maintain a large catfish population. Fish farmers can prepare for oxygen depletion by predicting oxygen consumption rates in their ponds and using emergency aeration equipment. Many farmers have permanent electrical aerators in their ponds that can be used daily (usually from midnight until dawn) during heavy feeding.

FEEDING PRACTICES

For maximum growth, the farmer is interested in a high rate of food consumption by the fish; however, because uneaten feed cannot be recovered, overfeeding represents not only an economic waste, but greater oxygen demand on the culture system. Knowledge of the amount, frequency, and methods of feed application under various production conditions are essential for the farmer to obtain maximum growth as well as feed efficiency.

Floating and Sinking Feeds. One of the most useful management tools in feeding catfish is the use of extruded or floating type feeds. Extruded fish feeds have advantages in that they allow the feeder to observe the fish feeding. By being able to see the fish eat, the feeder can feed the fish closer to their maximum rate of consumption without overfeeding, and disease and water quality problems can be detected more easily. Extruded feeds are also more water stable than pelleted feeds. However, their benefits must be evaluated against additional manufacturing cost when compared with compression pelleted (sinking) feeds. More energy is required to make extruded feeds and equipment cost is considerably higher.

Feeding sinking pellets can reduce feed cost by 10% to 20% when compared with feeding floating feeds. However, management skills must be better when feeding sinking pellets or the savings in feed cost will be offset by feed loss and water quality reduction. Sinking pellets must be fed over hard bottom areas so the feed will not sink in mud, and in areas free of rooted vegetation.

Some producers are successfully feeding both sinking and floating feeds in combination (85% sinking : 15% floating). They save 10% to 15% in feed costs and still have the management benefits of the floating feed.

Feeding Methods. On large commercial farms, feed is delivered from the plant in bulk form and stored in bins until fed from vehicle-mounted mechanical feeders (see Figure 7.2). In large ponds, feed is dispensed along all sides of the pond. The feed should be distributed over as large an area as possible to provide maximum feeding opportunity for the fish. This is necessary to provide uniform growth among the fish and, when there are various sizes of fish in the pond, to allow small fish opportunity to feed. When floating feeds are used and feeding activity of fish is slow or when strong winds prevail, feeding along the downwind side of the pond should be avoided.

Feeding Schedule. Traditionally, catfish farmers feed once daily, 6 or 7 days per week. However, research has shown that feeding twice daily, when water temperature is above 25° C, will allow for a 20% higher rate of feed consumption and a comparably faster rate of growth (Lovell 1979). The farmer must evaluate the benefits of faster weight

Fig. 7.2. Channel catfish ponds in the Mississippi River flood plain. The catfish are being fed from a feeder drawn by a tractor along the levee separating the two large (10-ha) ponds. Most catfish are fed an expanded floating pellet, enabling the farmer to gauge how much the fish consume.

gain in light of the additional cost of more frequent feeding. In addition to the extra cost, more frequent feeding increases the opportunity for wasting feed. This also means putting more feed into the pond, which increases the possibility for water quality problems. Feeding 7 days a week allowed for 17% more feed consumption and 19% more growth than feeding 6 days a week (Lovell 1979).

Optimum time of day for feeding catfish in ponds is influenced by dissolved oxygen and water temperature. Low dissolved oxygen (DO) depresses feeding activity of catfish. Figure 7.3 illustrates the diurnal variation in DO in intensively fed catfish ponds. Because of low DO in the pond water late at night and early in the morning, catfish should not be fed until well after sunrise and DO has risen to a level for active feeding by the fish. Also, feeding should not be done at night because the fish's increased oxygen requirement, which occurs in 4 to 8 hours subsequent to ingesting food, should not coincide with the decreased DO in the pond.

Summer water temperature in catfish ponds, especially shallow

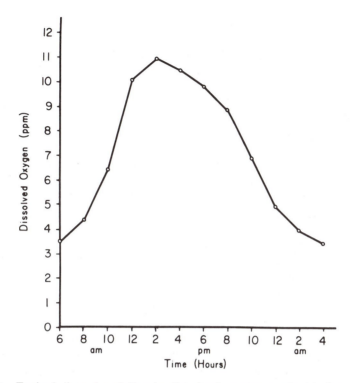

Fig. 7.3. Typical diurnal variation in dissolved oxygen content in intensively fed catfish ponds. Oxygen concentrations of 6–7 ppm represent normal water saturation and concentrations below 3 mg/L are stressful to the fish.

ponds with heavy algae blooms, often exceeds 32° C at the surface, which discourages feeding on floating feed. Under these temperature conditions the fish should be fed in the morning after the DO has risen, but before the water surface temperature has reached a stressful level. During spring and fall when seasonal temperatures are lower, the fish should be fed in afternoons when water is warmest.

Daily Feed Allowance.   Factors affecting the amount of feed channel catfish will eat are temperature, fish size, water quality, density of the pellet, energy content of the feed, the amount of feed consumed the previous day, and number of times fed per day.

The amount of feed that the pond can "metabolize" per day influences the amount of fish that can be produced. Dr. Homer Swingle, in his pioneer research on pond culture of catfish at Auburn University, found that approximately 32 kg of feed per hectare could be fed per day without serious risk of oxygen depletion. Later, with better understanding of pond dissolved oxygen dynamics and pond aeration equipment, catfish farmers have been able to increase this feeding rate up to three or four times. Thus, the daily feeding rate for growing catfish is determined by the amount of feed the fish will consume and the amount the farmer feels he or she can safely put into the pond.

Values in Table 7.1 represent daily feed allowances for channel catfish in the southern United States stocked in earthen ponds as 15-g fingerlings in the spring and fed to near appetite capacity for a 6-month growing season. The values were determined by feeding the fish to satiation on a floating commercial feed (35% protein and 3.1 kcal of digestible energy per gram) and removing and measuring all unconsumed feed put into the pond each day. Because factors such as temperature, atmospheric pressure, water chemistry and others affect day-to-day variation in food consumption by pond-raised catfish, the values in Table 7.1 represent 10% less than the amount of feed actually consumed by the fish to minimize waste.

A popular practice for catfish farmers is to keep a population of mixed sizes of fish in the pond at all times and periodically "harvest off" the larger fish. A readily available supply of fingerlings to replace the harvested fish is necessary. Under these conditions, some farmers keep a large population of fish in the ponds and feed a constant amount of feed throughout the growing season. Thus, the fish are usually underfed. The strategy here is to maximize feed efficiency by avoiding wasted feed.

Table 7.1. DAILY FEED ALLOWANCES FOR CHANNEL CATFISH IN PONDS IN THE SOUTHEASTERN UNITED STATES

Date	Water temperature[1] (° C)	Fish size (kg)	Feed allowance per day: Percentage of fish weight
4-15	20.0	0.02	2.0
4-30	22.2	0.03	2.5
5-15	25.5	0.05	2.8
5-30	26.7	0.07	3.0
6-15	28.3	0.10	3.0
6-30	28.9	0.13	3.0
7-15	29.4	0.16	2.8
7-30	29.4	0.19	2.5
8-15	30.0	0.27	2.2
8-30	30.0	0.34	1.8
9-15	28.3	0.40	1.6
9-30	26.1	0.46	1.4
10-15	22.8	0.50	1.1

[1] Mean temperature at 1 m depth at feeding time.
Source: Lovell 1977.

Feeding Young Catfish.   Newly hatched catfish fry live on nutriment from the egg yolk sac for 5 to 10 days, depending upon water temperature, after which time they will accept dry feed readily. Fry are usually held in tanks for a few days to a few weeks before being released into ponds. Feeds for tank feeding should be of small particle size (0.3–0.5 mm), high in protein (primarily from fish meal), and high in energy. It is essential that this feed be nutritionally complete. In fact, because of the small particle size, leaching of water soluble vitamins from the feed is a serious loss, so the feeds should be overfortified with water soluble vitamins. Fry in tanks should be fed several times daily. An automatic feeder is useful for this purpose. Water should flow generously through the tank to remove uneaten feed.

Transfer of the fry to ponds with a high zooplankton density as soon as possible is advantageous because catfish fry feed effectively on the small aquatic fauna. Supplemental feeding of fry in ponds should begin soon after stocking. A high-quality, 36% protein sinking feed, reduced from pellets to crumbles, is desirable for first feeding in ponds. As the fish increase in size to about 6 cm they can be fed floating feed of 5-mm diameter or continued on the crumbles (see Figure 7.4). For the first

Fig. 7.4.   Most catfish in the United States consume extruded (floating) feeds. Year 1 fingerlings, 6–16 cm in length, are fed either crumbles or small-size floating feeds (left); the second year, the "fingerling" size floating feed (center) is fed initially and the fish are finished on the "grower" size floating feed (right).

few weeks in the pond, this feed is actually a supplement; the small fish will get a large portion of their nutrients from natural pond organisms. As the fingerlings increase in size, they become more dependent on the supplemental feed.

Crumbles or pellets fed to fry and fingerlings should be kept to the maximum size that the fish can ingest to minimize dissolving of ingredients in the water. Optimum feed particle sizes for small catfish are shown in Table 7.2.

Winter Feeding.   The optimum temperature for catfish growth is approximately 30° C; as temperature decreases, food consumption decreases. Generally, catfish do not feed consistently in ponds when the water temperature drops below 21° C; although catfish will feed at temperatures as low as 10° C, the amount and frequency of feeding are reduced. Advantages of winter feeding of channel catfish are that skillful feeding can effect a weight gain or at least prevent a weight loss, and it can keep fish healthier throughout the winter. With unskilled feeding, however, feed can easily be wasted and accumulation of a large amount of unconsumed or undigested feed in the pond during the winter may cause water quality problems in the spring as this feed begins to oxidize when the water warms up.

Catfish held in ponds during winter will lose weight if not fed. A study at Auburn University showed that marketable-size (0.45-kg) catfish held overwinter in ponds (November 15–March 15) without feeding lost 9% of their weight, while those fed 1% of body weight on days when water temperature was above 12° C gained 18%.

Winter temperatures significantly influence the effects of feeding or not feeding of catfish. Studies in Mississippi showed that fed channel catfish fingerlings made significantly more growth during a mild winter than during a cold winter (mean daytime temperatures of water were 13° C and 7° C). During the mild winter, fish fed 1% of their weight three times a week gained 21%, while those fed at the same rate six times a week gained 45%. During the cold winter, the fish that

Table 7.2. OPTIMUM FEED PARTICLE SIZES FOR SMALL CATFISH

| Fish size (cm) | Feed particles | | Feed allowance per day (% of fish wt.) |
	Size number	Max. diameter (mm)	
Swim-up fry	Starter	0.5	6–10
1.5–4	#1	0.8	6
4–6.5	#2	2.0	5
5–16	#3	3.3	3–4

were fed three times a week gained 24% and those fed six times a week gained only 29%. Catfish lose more weight during mild winters than during cold winters. Studies at Auburn University showed that mean weight loss by marketable-size catfish during a cold winter was only 6.3% as compared to 9% during a milder winter. During the colder winter, the weight loss was all fat; during the milder winter there was some protein loss also.

Fish accustomed to floating feeds will feed on the surface during winter, but more slowly than during summer. A sinking pellet or a "slow-sink" extruded feed can be fed in winter. A recommended guide for winter weather feeding of catfish in ponds is to feed food-size fish around 0.75% of their estimated weight when water temperature at 1 m (3 ft.) depth is $\geq 13°$ C. Fingerling fish can be fed 1% of body weight three times per week or daily with extended periods of warm weather.

Food size catfish respond as well to low protein (25%) feeds as to high protein (35%) feeds in cool weather, thus 25% protein diets are recommended for winter feeding of marketable-size fish (> 0.25 kg). Fingerlings can be fed 32% to 36% protein feeds.

NUTRITIONAL REQUIREMENTS OF CHANNEL CATFISH

Nutritional requirements of channel catfish have been defined well enough that nutritious, functional, and economical diets are available to the catfish farming industry. Still needed, however, is more information on effects of fish size on nutrient and energy requirements and physiological availability of nutrients and energy from various sources.

Essential amino acid requirements of channel catfish are presented in chapter 2. Cystine can replace or spare about 60% of methionine on a molar sulfur basis. Tyrosine can spare about 50% of the total phenylalanine requirement for channel catfish. Methionine plus cystine appear to be the first limiting amino acids, followed by lysine.

Total protein requirements for optimum growth by channel catfish have been reported to vary from 25% to 36%. Major reasons for these differences in response to varying dietary protein percentages are variations in fish size, daily feed allowance, amount of nonprotein energy in the diet, protein quality, water temperature, and amount of natural food available in ponds.

Page and Andrews (1973) found that small catfish increased weight gain more when dietary protein level was increased from 25% to 35% than did larger fish in raceways, although data have been presented (chapter 2) that the protein/energy ratio does not change much as size increases up to 250 g. At present, very little information is available for recommendations on decreasing protein allowance in channel catfish feeds as size increases. There may be reluctance in the catfish farming industry to practice phase feeding where the dietary protein percentage changes as fish size increases because many farmers culture a range of sizes of fish in the ponds simultaneously and periodically "harvest off" the larger ones and replace them with small fish.

Coldwater and warmwater fish respond to higher protein levels at higher water temperatures. An experiment at the Fish Farming Experimental Station in Stuttgart, Arkansas, demonstrated that at temperatures below 24° C, channel catfish in ponds grew no better on 35% than on 25% protein feeds. However, when water temperatures exceeded 24° C, the fish gained more on the higher protein feeds. Food-size catfish overwinter on 25% protein diets as well as on higher protein diets.

Energy levels in catfish feeds have been treated casually because a deficiency or excess of energy will not affect the health of fish appreciably, and practical feeds made with commonly available ingredients are not likely to be extremely high or low in energy. Nonetheless, the optimum energy requirement in commercial catfish feeds is important because a deficiency of energy in the ration means that protein and other nutrients will not be utilized to their fullest potential for growth, and too much energy in the ration can limit the daily intake of food and can cause excessive fat deposition and reduce dress-out percentage.

Data from Table 2.1 in chapter 2 (page 13) show that the optimum ratio of digestible energy to crude protein for maximum protein gain by channel catfish of 3-g to 250-g size was 10 kcal to 11 kcal per g of protein. Commercial catfish feeds containing 32% crude protein contain about 9 kcal/g of protein. Evidence has been presented that reducing the energy-to-protein ratio below 9 kcal per g of protein will increase dress-out yield by reducing the amount of abdominal fat and skin fat in the fish. However, the only way commercial catfish feeds can significantly reduce the energy-to-protein ratio without using a nonnutritive ingredient is to increase protein percentage in the diet, which may be economical under some cost-price conditions.

Channel catfish digest the energy in animal feedstuffs (fish meal, meat, and bone meal) which comes mainly from protein and fat, as well as do livestock (80–85% digested). Energy from fat-extracted oilseed meals (soybean, cottonseed), which comes partially from carbohydrates and partially from protein, is less digestible to catfish (53–59%

digested). Energy in grains, which comes primarily from starch, is also less digestible to fish than to livestock (50–65% digested). Cooking, such as in extrusion processing of fish feeds, increases digestibility of starch to channel catfish by 10% to 15%.

In the early years of catfish farming, commercial feeds contained no vitamin supplements. Signs of nutritional deficiency were seldom observed because fish density in the ponds was low (1,000–2,000 kg/hectare) and natural pond foods were a source of vitamins. However, Prather and Lovell (1972) added a vitamin premix designed for poultry to catfish feeds and increased weight gain by 19% in ponds. As stocking density in ponds began to increase and as other culture methods were tried where natural foods were not available, the nutrient composition of commercial feeds became more critical. A classical example is the discovery of the "broken back syndrome" in catfish by pathologists in the early 1970s. Moribund, deformed fish were found with no infectious disease organisms. These fish came from culture systems where natural food was scarce and had been fed feeds that at that time were designed as supplemental feeds for use in ponds. Research showed the absence of vitamin C in the feed was responsible for this condition. Pond feeding experiments showed that as the standing crop of fish increased above 3,000 kg/hectare, vitamin C deficiency would occur in culture ponds. Today catfish feeds used by commercial fish farmers in the United States are nutritionally complete.

The vitamin requirements for channel catfish are presented in Table 2.6 in chapter 2 (page 33). These values do not allow for processing or storage losses. Inositol and biotin are usually not added to commercial catfish feeds; these vitamins are found in sufficient quantity in the feed ingredients.

Several studies have shown that channel catfish and rainbow trout fed diets containing higher than the normal requirements for vitamin C have increased resistance to certain bacterial infections. Durve and Lovell (1982) demonstrated that resistance of small channel catfish to infection from *Edwardsiella tarda* was enhanced when the dietary level of vitamin C was increased to five times the requirement for normal growth. Later, Li and Lovell (1984) found that dietary doses of vitamin C much above the minimum requirement for normal growth, increased resistance to infection with *E. ictaluri* and increased specific and non-specific immune responses in small channel catfish in aquariums.

Phosphorus is of major importance in formulation of practical catfish feeds because of the relatively high physiological requirement, the low amount of available phosphorus in feedstuffs of plant origin, and the relatively low level of dissolved phosphorus in natural waters. The physiological requirement for calcium is also high; however, channel catfish absorb a significant amount from the water. Minimum requirement for available phosphorus in diets for rapidly growing channel catfish is approximately 0.45%, whereas dietary calcium is usually not considered in feed formulation.

Trace minerals are sometimes deficient in plant ingredients produced in mineral deficient areas, thus fish feeds containing low levels of animal byproducts may be deficient in one or more trace minerals. A premix to provide the requirements of zinc, iron, manganese, iodine, copper, and selenium similar to that presented in Table 7.3 (footnote) is recommended in commercial feeds.

A model formulation for a catfish feed similar to that fed on commercial farms to grow fish from fingerling to harvestable size is given in Table 7.3. The vitamin and trace mineral premixes footnoted in the table are sufficient to make the feed nutritionally complete.

IMPORTANCE OF NATURAL FOODS IN CATFISH PONDS

The nutrient and energy contribution from natural food organisms to channel catfish in intensively stocked and fed ponds is very small. Wiang (1977) made an estimate of this by stocking fingerling catfish in nine ponds (7,500 fish/hectare) and subjecting them to three treat-

Table 7.3. MODEL FORMULA FOR A PRACTICAL EXTRUDED CATFISH FEED

Ingredient	International feed number	Amount (%)
Menhaden fishmeal	5-02-009	8.0
Soybean meal (48% protein)	5-04-612	48.2
Grain or grain by-products		41.0
Dicalcium phosphate		1.0
Fat (sprayed on after pelleting)		1.5
Trace mineral mix[1]		0.05
Vitamin mix[2]		0.25
Vitamin C		0.025

Note: If the formula is pelleted instead of extruded, use 2.5% organic pellet binder in place of equal amount of grain.
[1] Mineral mix should provide the following minerals in approximately these amounts in mg/kg of feed: manganese, 5; zinc, 100; iron, 40; copper, 5; iodine, 5; cobalt, 0.05; and selenium, 0.3.
[2] Vitamin mix should provide the following vitamin activities in approximately these amounts per kg of feed: vitamin A, 3,000 IU; vitamin D_3, 1,000 IU; vitamin E, 50 mg; menadione, 10 mg; choline, 500 mg; niacin, 80 mg; riboflavin, 12 mg; pyridoxine, 10 mg; thiamine, 10 mg; pantothenic acid, 32 mg; folic acid, 5 mg; and vitamin B_{12}, 8 μg.

ments: three ponds received supplemental feed; three ponds contained fish confined in cages that were fed the supplemental feeds (see Figure 7.5); and three ponds were "enriched" with organic and inorganic nutrients, in a liquid suspension so the fish could not eat them, in amounts equal to the wasted nutrients in the fed ponds. It was assumed that the fertility of the ponds in treatment 3 would be nearly the same as that of the fed ponds. The fish fed in cages gained 9% less than those fed in the open pond. The difference was attributed to the availability of pond food to the fish in open ponds. The fish in the unfed but "enriched" ponds gained 8.3% as much as those in the open ponds. Thus, the growth by channel catfish from natural pond food appeared to be only 8.3% to 9% of the total gain in intensively fed ponds.

EFFECT OF FISH SIZE ON FEEDING RESPONSES

There are several economic reasons why catfish farmers should know the feeding responses of fish of various sizes. If farmers have a choice of harvesting the fish at a given size or waiting until the fish grow larger,

Fig. 7.5. Feeding channel catfish in suspended cages. Cage cultured catfish require nutritionally complete feeds because they are deprived of nutrients from pond organisms.
(Courtesy of H. R. Schmittou)

they would want to anticipate the growth rate, feed consumption, and change in feed efficiency if they fed the fish to a larger size. It is generally assumed that growth rate and feed efficiency decrease as size increases in fish as in warmblooded food animals; however, this type of information for fish is scarce.

A study was conducted at Auburn University to obtain feeding response information on various sizes of channel catfish under pond culture conditions. A 3-month feeding study was conducted during the middle part of the growing season (June through August) when water temperature and quality were relatively constant and favorable for rapid growth. The results (Table 7.4) show that the smaller fish (45 g initial wt.) consumed more feed, grew faster, and had better feed conversion rates than larger fish (150 and 550 g initial wt.). Generally, as fish size increased, feed consumption, growth rate and feed efficiency decreased. The study also revealed that the smallest groups of fish fed more consistently and offered less opportunity for feed waste, and were more uniform in size at harvest than the larger fish.

COMPENSATORY GROWTH IN CHANNEL CATFISH

Compensatory growth in animals represents increased growth rate as a result of growth restriction caused by food deprivation during an earlier period. The ensuing catch-up following restricted feeding is characterized by an increase in efficiency of weight gain in most animals. This is claimed to be caused by the "filling out" of existing body cells.

Does this suggest that pond-raised channel catfish in temperate

Table 7.4. EFFECT OF SIZE OF CHANNEL CATFISH ON FEED CONVERSION, GROWTH RATE, AND FEED CONSUMPTION FOR THREE GROWTH PERIODS IN EARTHEN PONDS

Criterion	Initial fish size (g)	June	July	August	Overall
Feed conversion	45	1.25	1.35	1.51	1.43
(Kg feed/kg gain)	150	1.27	1.58	1.82	1.67
	550	1.82	2.39	2.08	2.16
Growth increase	45	229	391	688	
(Percentage of	150	202	274	405	
initial weight)	550	118	134	169	
Daily feed consumption	45	3.0	2.4	1.6	
(Percentage of	150	2.6	1.6	1.5	
body weight)	550	1.2	1.4	1.2	

zones not fed during the winter will perform as well when fed the following spring and summer as fish that have been fed during the winter? It is well known that pond-raised fish in temperate zones feed voraciously in the spring following a winter of limited feeding. A study conducted in Arkansas showed that winter-fed channel catfish, which received feed on days when water temperature was $\geq 10°$ C, gained 9.4% during the winter, while a companion group that was not fed lost 1.2% of their weight. At the end of the following summer growing season, the fish not winter-fed had increased their size by 547% as compared to 479% for the winter-fed fish. The study suggests that channel catfish can experience compensatory growth; however, before winter feeding of catfish is discontinued, additional information is needed on the effects of winter feed deprivation on fish condition and health.

EFFECTS OF FEED ON SENSORY QUALITIES OF PROCESSED CATFISH

Flavor.   Commercial feeds usually have very little effect on fish flavor. Grains, oilseed meals, and animal byproduct meals generally do not impart characteristic sensory qualities. Fats or fat-soluble materials are the most likely to affect taste, appearance, or keeping quality (frozen) of fish flesh. Diets containing high levels of marine fish oil will impart "fishy" flavor in the fed fish. This is usually undesirable in freshwater species, such as channel catfish.

Appearance.   Some fish in the wild have brightly pigmented flesh and others have white flesh. The consumer usually demands the same appearance in cultured fish as is found in that species in the wild. Cultured salmon must have pink flesh or receive a low market price. Conversely, channel catfish flesh is preferred white. Channel catfish will concentrate yellow pigment in the anterior, dorsal area of the muscle. These pigments are xanthophylls (leutin and zeoxanthin) and the concentration in the flesh is influenced by concentration in the diet. This pigmented area is undesirable in catfish and feed processors are careful to minimize the use of xanthophyll-containing ingredients in catfish feeds. Over 11 mg of xanthophylls/kg feed will impart yellow color in the flesh. Blue-green algae is also a source of the pigment.

Fattiness in Cultured Catfish.   Fish cultured for release into natural waters, like hatchery-reared trout and salmon, benefit from ample stores of body fat in higher survival rates. Conversely, fattiness is generally undesirable in fish cultured for food. In the United States, consumers relate fish to lean muscle with firm, flaky texture and discriminate against greasy, soft texture. Fattiness reduces yield in processed fish. Also, fatty fish are more susceptible to oxidative rancidity and do not hold up well in frozen storage.

Usually, fish caught from the wild are less fatty than intensively fed cultured fish. One reason why this might be expected is heavy feeding rate. Another reason is that the ratio of energy to protein is higher in prepared diets than in natural foods. A typical practical catfish diet contains 8 kcal to 10 kcal of digestible energy per gram of protein, while the energy/protein ratios in aquatic fauna (insects, larvae, crustaceans) are in the order of 5 to 6.

The average fat percentage in cultured catfish at present is higher than a decade ago. There are several reasons for this. One is heavier feeding rates. In the early years, catfish farmers fed conservatively, especially during the latter part of the growing season, but with better knowledge or water quality and pond management, feeding is less restricted and catfish are fed heavily up to harvest. Processing larger size fish also accounts for an increase in fattiness. Due to changing demand for larger size fish, catfish stay in production ponds longer and are harvested at sizes of 1 kg and larger as compared to an average harvest size of around 0.45 kg in early years. Data in Table 7.5 are from a feeding study in which fish of three sizes were grown simultaneously in the same ponds and fed at either restricted or satiation rates. The data indicate that larger catfish have more fat in the body after processing. The data also show that the feeding rate only affected body fat in the smallest size fish (one-third of the stocked population). This may indicate that the restricted feeding regime only limited the feed intake of the smaller fish, while the larger fish consumed feed at near satiation rate. Increasing the ratio of energy to protein in feed will also increase body fat percentage. Data in Table 7.6 show that increasing protein percentage, which reduced the energy to protein ratio, reduced body fat and increased dressing percentage in the fed fish.

Table 7.5. FAT PERCENTAGE IN DRESSED CHANNEL CAT-FISH OF VARIOUS SIZES FED TO SATIATION OR AT A RESTRICTED RATE

Fish size (kg)	Satiation fed	Restricted fed
0.22	9.5	7.9
0.55	11.2	11.3
1.10	12.2	12.0

Source: Cacho (1984).

Table 7.6. EFFECT OF DIETARY ENERGY-PROTEIN RATIO (DE/P) ON WEIGHT AND FAT GAINS, FAT PERCENTAGE, AND DRESSING PERCENTAGE IN CHANNEL CATFISH

Dietary protein percentage	DE/P	Weight gain per fish (g)	Fat gain per fish (g)	Percentage fat		Dressing percentage
				Dressed fish	Offal	
26	9.9/1	254	41	11	18	54.5
32	8.5/1	258	33	9	14	56.6

Source: Deru (1985).

FEEDING BROOD FISH

Early research showed that feeding artificial (pelleted) diets to brood channel catfish in ponds resulted in poor spawning success and that supplementation with live or fresh fish or animal parts was necessary for successful spawning. However, these early artificial diets were not nutritionally complete and, because brood catfish 2 kg to 5 kg in size do not obtain much natural food from ponds without forage fish, the nutrients supplied by the supplemental fish and animal tissues were necessary. Many catfish farmers continue to supplement commercial feeds for brood catfish with forage fish.

Results from a 2-year pond study at Auburn University showed that growth and reproductive characteristics (number of eggs, size of eggs, number of spawns, spawning date, hatching rate, and quality of the fry) of channel catfish females fed a nutritionally balanced, pelleted diet alone were equal to those of fish fed the pelleted diet plus frozen golden shiners. Females fed a pelleted diet without a vitamin supplement or golden shiners spawned later and had fewer and smaller eggs than females fed the nutritionally adequate pelleted diet. When the vitamin-deficient pelleted diet was supplemented with golden shiners, reproductive performance was equal to that of fish fed the high quality diets. The fish fed only frozen golden shiners performed as well as fish fed the frozen fish plus a pelleted diet. Male channel catfish fed the pelleted diet without vitamin supplementation in ponds had low quality spermatozoa, while those fed the nutritionally complete pelleted diet had normal spermatozoa.

The pelleted diets used in this study were made from commercial feedstuffs. The diet without vitamin supplementation was devoid of vitamin C and probably deficient in vitamins A, D, E, pantothenic acid, riboflavin, and possibly some of the other water soluble vitamins. Results of this research indicate that diet quality has an important effect on reproductive performance of channel catfish and that live or frozen raw fish is not a necessary part of the diet when a nutritionally balanced, pelleted feed is used.

REFERENCES

CACHO, O. J. 1984. Effects of fish size and dietary protein level on growth, feed consumption and feed efficiency by channel catfish. Master's thesis, Auburn University, Auburn, AL.

DERU, J. 1985. Effects of dietary energy-protein ratio on composition of gain by channel catfish and autoxidation of the fish in frozen storage. Master's thesis, Auburn University, Auburn, AL.

DURVE, V. S., and R. T. LOVELL. 1982. Vitamin C and disease resistance in channel catfish. *Can. J. Fish. Aquat. Sci.* 39: 948–951.

LI, Y., and R. T. LOVELL. 1984. Elevated levels of dietary ascorbic acid increase immune responses in channel catfish. *J. Nutr.* 115: 123–131.

LOVELL, R. T. 1977. Feeding practice. In Nutrition and feeding of channel catfish, eds. R. R. Stickney, and R. T. Lovell. *South. Coop. Ser. Bull.* 218.

LOVELL, R. T. 1979. Factors affecting voluntary food consumption by channel catfish. *Proceedings of the World Symposium on Finfish Nutrition and Fish Feed Technology,* pp. 556–561. Hamburg, 20–23 June, 1978.

PAGE, J. W., and J. W. ANDREWS. 1973. Interactions of dietary levels of protein and energy on channel catfish (*Ictalurus punctatus*). *J. Nutr.* 103: 1339–1346.

PRATHER, E. E., and R. T. LOVELL. 1972. Effect of vitamin fortification in Auburn No. 2 fish feed. *Proc. S.E. Assoc. Game and Fish Comm.* 25: 479–483.

United States Department of Agriculture. 1987. Catfish market report. SpCr 8. Washington, DC: Agricultural Statistics Board.

WIANG, C. 1977. Nutritional contribution of natural pond organisms to channel catfish growth in intensively-fed ponds. Ph.D. diss., Auburn University, Auburn, AL.

8

Practical Feeding—Tilapias

Chhorn Lim

Tilapias are endemic to Africa, but are presently found in most tropical and subtropical regions of the world. They have become a top priority fish for culture in the tropics because of their fast growth, efficient use of natural aquatic foods, propensity to consume a variety of supplemental feeds, herbivorous nature, resistance to diseases and handling, ease of reproduction in captivity, and tolerance to wide ranges of environmental conditions. Although indigenous to fresh water, tilapias are euryhaline and able to survive, grow, and some species even reproduce in seawater up to 40 mg/ml (ppt) salinity. Some of the cultured species have been shown to survive dissolved oxygen concentrations of 0.1 mg/L. They grow over a pH range of 5 to 11 and tolerate an unionized ammonia concentration of 2.4 mg/L. However, tilapias are not able to survive a water temperature below approximately 8° C to 12° C. Their activity and feeding become reduced below 20° C and feeding stops around 16°C.

Most cultured tilapias are grouped into two genera (Trewavas 1982): *Tilapia,* which are macrophagous and substrate-spawners; and *Oreochromis,* which are microphagous and mouth-brooders. About 70 species have been identified under these two genera; however, only two *Tilapia* species, *rendalli* and *zillii,* and three *Oreochromis* species, *mossambicus, niloticus,* and *aureus,* have been used widely in practical culture. For practical purposes, and because most of the publications on tilapias still carry the old generic name, the common name of both genera is referred to here as *tilapia* (see Figure 8.1).

CULTURE PRACTICES

Seed production.   Unlike most fish species, in which spawning is one of the major culture problems, minimum skill and experience are required to spawn tilapias. Most tilapias are able to reproduce at 5 to 6

Fig. 8.1.   Nile tilapia, approximately 0.4 kg in size.

months of age and can spawn every 6 to 8 weeks at water temperatures between 25° C and 32° C. The total number of eggs produced per spawning is small and differs among species and size of fish. The mouth-brooding species lay fewer eggs than the substrate spawners. A large-size substrate spawner can lay as many as 7,000 eggs per spawning, whereas the mouth-brooders seldom produce more than 2,000 eggs. Although the fecundity of tilapias is low, the high frequency of breeding and the high rate of larval survival often create problems of overpopulation and stunting which result in most of the fish not reaching marketable size. To overcome this problem, it is necessary to use fast growing species that reach marketable size before they breed, such as *O. aureus* and *O. niloticus,* or to practice monosex (all-male) culture or polyculture with a suitable predator fish.

In monosex culture, males are preferred because they have a faster growth rate than females. Monosex culture is managed by manual sexing, hybridization, or sex-reversal of genotypic females with the use of hormones. Manual sexing is done by selecting the males after the fingerlings have reached 20g to 50g and have well developed sexual parts. Hybridization between certain species of tilapia can produce a high percentage of males (85–100%). However, a disadvantage of this technique is the difficulty in maintaining pure stocks that produce a high percentage of males. Sex reversal to produce monosex male populations can be accomplished by administering androgenic hormones during early larval stage by means of injection, submerging the fry in an aqueous hormone solution, or incorporating hormones into the diets. The latter is the most convenient and effective method. Methyltestosterone or ethynyltestosterone is incorporated at a concentration of 30 mg/kg to 60 mg/kg of diet and fry are fed at 10% to 12% of their body weight per day, divided into three or four feedings, for approximately 4 weeks. This method of monosexing of tilapias for culture is practiced commercially in several areas of the world.

Breeding of tilapias can be done in earthen ponds, nets, or tanks. Mature fish are stocked at a ratio of 2 to 5 females to 1 male. If females and males are of different species, the ratio may be as low as 1 to 1. The stocking density varies from 1 to 4 fish/m^2. Broodfish should be fed and those spawned in an artificial environment should receive a nutritionally complete feed. At water temperatures of 25° C to 30° C, fry can be found in about 10 days to 14 days after the breeders are stocked. The fry should be removed from spawning facilities at regular intervals, usually weekly or biweekly, by use of a 6-mm^2 mesh seine. The fry collected from the spawning areas are usually transferred to nursery ponds or tanks for rearing to a size of 20 g to 50 g for subsequent stocking into production ponds.

Culture methods. There is great diversity among culture systems and husbandry methods being used for producing marketable size tilapias. Pond culture is the most commonly used system (see Figure 8.2). Stocking rates vary depending on the size of fish, type and level of nutrient inputs, culture period, rate of water exchange, aeration, and other management practices. Small-scale farmers in areas where commercial diets are unavailable or expensive often use only manures (animal excreta or compost) or inorganic fertilizers. When natural food constitutes an important source of nutrients, supplementary feeding with locally available, inexpensive feed materials, such as rice bran, copra meal, brewery waste, coffee pulp, and similar materials, can increase production appreciably. As stocking rate increases, the natural food becomes less significant and better quality supplemental feeds are needed. The growing period may last from 3 months to 6 months, depending on preferred market size and management practice. Yield figures for various tilapia monoculture production systems are summarized in Table 8.1.

Fig. 8.2. Tilapia culture ponds (1 hectare size) in Jamaica. Ponds are fertilized with organic or inorganic fertilizer to enhance production of natural foods and fish are fed a 25% protein pelleted feed. Two crops per year are produced; yield is approximately 5,000 kg/hectare/crop.

Polyculture systems for tilapias and other species with different food habits are used to maximize the utilization of available natural foods. In Israel, tilapias are stocked in combination with common carp at a 60:40 ratio; sometimes silver carp are added. A practice in Taiwan is to stock 12,000 to 15,000 tilapia/hectare together with 600 bighead carp, 600 silver carp, 300 grass carp, and 100 common carp in ponds provided with heavy fertilization and single-ingredient supplemental feeds. Integrated farming of tilapia with poultry or pigs is practiced in many developing countries. The manures of fed chicken, ducks, or pigs are used to fertilize the ponds or to serve directly as food for the fish. Rice-fish farming is another type of integrated farming, which has long been practiced in Southeast Asia. This system involves rearing of fish either simultaneously with the growing rice crop or on a rotational basis. Management conditions and yield data for various polyculture and integrated farming systems with tilapias are summarized in Table 8.2.

Intensive culture of tilapia has gained popularity in recent years. Fish are stocked at very high densities in tanks (see Figure 8.3),

Fig. 8.3. Tilapia are adaptable to intensive culture systems. Here, they are fed nutritionally complete diets and grown successfully in circular raceways supplied with geothermal well water in the western United States.

raceways, or earthen ponds with flowing water and aeration, or in floating net cages, and fed with high quality pelleted feeds. Production data for various intensive culture practices are presented in Table 8.2.

NUTRIENT REQUIREMENTS

Proteins and amino acids.   The minimum dietary level of an amino acid–balanced protein required for optimum growth in absence of natural food is near 50% for tilapia fry and decreases to about 35% as fish increase to 30 g in size (Table 8.3). Practical requirements for larger fish have been reported from 25% to 35% of the feed; this varies with size of the fish, amount of natural food in the culture system, and dietary factors such as quality of protein and energy level.

Tilapias require the same 10 essential amino acids as other fishes and land animals. These are arginine, histidine, isoleucine, leucine, lysine, methionine, phenylalanine, threonine, tryptophan, and valine. The quantitative requirements for these essential amino acids for growth by young *O. niloticus* (Nile tilapia) are presented in Table 2.4

Table 8.1. PRODUCTION OF TILAPIAS IN MONOCULTURES

Culture system	Culture facilities	Sex/Species	Stocking rate (1,000/ha)	Major inputs	Culture period (days)	Net production (kg/ha)	Reference
Extensive	Ponds	All-male hybrids (*O. niloticus* ♀ × *O. hornorum* ♂)	5.6	None	253	288	Lovshin 1982
"	"	"	10.0	Fresh cattle manure daily; total manure used: 28,381 kg/ha	103	1,646	Collis and Smitherman 1978
"	"	"	8.0	Chicken manure, 500/kg/ha/week	189	1,150	Lovshin 1982
"	"	"	8.96	Fresh cattle manure, 840 kg/week	253	927	"
Semi-intensive	Ponds	"	10.0	Triple superphosphate, 56 kg/15 days; Ammonium sulfate, 56 kg/15 days; Rice bran, 3% body wt/day	180	1,648	"
"	"	"	10.0	Floating catfish feed (36% protein), 3% body wt/day	103	2,663	Collis and Smitherman 1978
"	"	"	11.25	Castor bean meal, 3% body wt/day	238	3,276	Lovshin 1982
"	"	"	11.25	Palm nut cake, 3% body wt/day	238	2,447	"
"	"	"	11.25	Cottonseed cake, 3% body wt/day	238	2,095	"
"	"	"	25.0	Triple superphosphate, 540 kg/ha/yr; Ammonium sulfate, 540 kg/ha/yr; 50% palm nut cake, 50% cottonseed cake, 5% body wt/day	367	10,000	"

System	Culture system	Species/sex	Stocking	Feeding	Culture period (days)	Production	Reference
	"	All-male O. niloticus	15.0	Fed with supplemental diet	70–105	1,800–2,900	Popma et al. 1984
	"	All-male O. aureus	30.0	Fed with a 25% protein diet (10% fish meal)	76	4,890	Tal and Ziv 1978
Intensive	Ponds	"	30.0–50.0	Fed with rich feeds and provided with aeration (paddle wheels)	2 crops/yr	27–45,000/yr	Liao and Chen 1983
	"	"	80.0	Fed with a 25% protein diet (10% fish meal) and provided with aeration	100	16,750	Tal and Ziv 1978
	"	Hybrids (O. niloticus ♂ × O. mossambicus ♀)	59.0	Fed with 34% protein diet, 3% body wt/day	150	8,014	Sin and Chiu 1983
	"	"	111.0	Fed with 34% protein diet, 6% body wt/day, provided with aeration	72	9,740	"
	Concrete tanks	Mixed sex O. aureus	200.0	Fed with 35% protein diet, 5 to 2% body wt/day, provided with aeration	114	17,307	Allison et al. 1976
	"	All-male hybrids (O. niloticus ♀ × O. aureus ♂)	500.0–1000.0	Fed with rich feeds, 3–4 times daily, changed part of water twice daily, water recirculated and provided with aeration	90–150	300–400,000	Liao and Chen 1983
	Floating cages (7 × 7 × 2.5 m)	"	800.0–1000.0	Fed protein-rich feed 3 times daily	120–180	440–550,000	"

Table 8.2. FISH PRODUCTION IN INTEGRATED AND POLYCULTURE SYSTEMS

Culture system	Fish/ha	Fish species	Stocking rate (1,000/ha)	Major inputs	Culture period (days)	Net production (kg/ha)	Reference
Rice-fish		Mixed sex *O. niloticus*	3–5	150 kg/ha of 16-20-0 and 75 kg/ha of urea	70–138	94–220	dela Cruz 1980
Pig-fish	60	*Cyprinus carpio* *Ophicephalus striatus*	17 2 0.2	Only pigs were fed. Manure was washed into the ponds daily.	90	1,576 353 21	Cruz and Shehadeh 1980
"	210	Tilapia hybrids (*O. niloticus* ♂ × *O. mossambicus* ♀) Grass carp Bighead carp Silver carp Mullet Common carp Sea perch	30 0.3 0.4 1.5 1.5 1.5 0.3		210–280	4,560 340 576 765 315 612 203	Chen and Li 1980
Duck-fish	1,250	Mixed sex *O. niloticus* *Cyprinus carpio* *Ophicephalus striatus*	17 2.8 0.2	Only ducks were fed. Duck droppings were collected daily and broadcast into the pond.	90	679 291 44	Cruz and Shehadeh 1980

		Species		Feeding			Reference
"	2,200	Tilapia hybrids (*O. niloticus* ♂ × *O. mossambicus* ♀)	4.7		240–300	2,900	Chen and Li 1980
		Grass carp	0.3			324	
		Bighead carp	0.3			382	
		Silver carp	1.2			864	
		Mullet	3			495	
		Sea perch	0.2			144	
		Common carp	1			540	
		Eel	0.04			5	
		Clarius catfish	0.07			17	
"	70	All-male hybrids (*O. niloticus* ♀ × *O. hornorum* ♂)	8	Only pigs were fed. They received 5% of their weight daily of a compound feed.	189	1,290	Lovshin, 1982
"	60	"	10	Pigs were fed 5% of their body weight with a compounded feed. Fish were fed 2% of their body weight daily, 6 days/week.	193	2,733	"
"	120	"	10	Pigs were fed 5% of their body weight with a compounded feed.	180	2,510	"

171

Table 8.2. (continued)

Culture system	Fish/ha	Fish species	Stocking rate (1,000/ha)	Major inputs	Culture period (days)	Net production (kg/ha)	Reference
Polyculture		All-male hybrids (O = niloticus ♀ × O = hornorum ♂)	5	Fed with chicken pellet 3% of the body weight of C. macropomum per day, 6 days/week	360	3,267 7,359	"
		Colossoma macropomum	10				
"		Mixed sex O. aureus Common carp Silver carp Grass carp	5 7.1 1 1.4	Fish were fed with grain only and changed to 25% protein feed when the standing crop exceeded 800 kg/ha. Ponds were fertilized weekly with 60 kg superphosphate and 40 kg chicken manure/ha	140	1,018 1,776 1,576 180	Halevy, 1979

Table 8.3. DIETARY PROTEIN REQUIREMENTS FOR MAXIMUM GROWTH
BY TILAPIAS

Species	Size range (g)	Protein source	Protein requirement (%)
O. aureus	Fry–2.5	Casein and egg albumin	56
	2.5–7.5	Casein and egg albumin	34
O. mossambicus	Fry	Fish meal	50
	0.5–1.0	Fish meal	40
	6–30	Fish meal	30–35
O. niloticus	1.5–7.5	Casein and gelatin	36

Sources: O. aureus, Winfree and Stickney (1981); *O. mossambicus*, Jauncey and Ross (1982); *O. niloticus*, Kubaryk (1980).

in chapter 2 (page 26). The requirement for the sulfur amino acids, methionine and cystine, can presumably be met by either methionine alone or a combination of methionine and cystine. Dietary cystine can substitute up to 50% of the total sulfur amino acid requirement for *O. mossambicus* (Jauncey and Ross 1982).

Casein has good protein quality for *O. aureus*, followed in decreasing order by fish meal, soybean meal, peanut cake, and yeast (Wu and Jan 1977). Fish meal is a better protein source than plant proteins from copra, groundnut, soybean, sunflower seed, rapeseed, and cottonseed for *O. mossambicus* (Jackson, Capper, and Matty 1982). When the 10 essential amino acids were individually added to an all-soybean protein diet, only histidine, isoleucine, phenylalanine, and valine could increase growth of *O. aureus* (Wu and Jan 1977). However, supplementation of the limiting essential amino acids is not required when soybean protein is supplemented with fish meal (Viola and Arieli 1983).

Tilapias digest the protein of fish meal and meat and bone meal well, equal to channel catfish. However, the digestibility of the protein in cereal grains and oilseed meals is higher for tilapias than for channel catfish (Popma 1982). The digestibility of protein of some common feed ingredients by *O. niloticus* is given in Table 8.4.

Energy.   The dietary protein to energy ratio required for maximum growth decreases with increasing size of tilapia. Winfree and Stickney (1981) found that small *O. aureus* (2.5 g) grew best when the diet contained 56% protein with a digestible energy/protein (DE/P) ratio of 8.2 kcal/g of protein. Larger fish (7.5 g) grew maximally when fed a

Table 8.4. DIGESTIBILITY COEFFICIENTS FOR PROTEIN, FAT, CARBOHYDRATE, AND GROSS ENERGY IN FEED INGREDIENTS BY *O. NILOTICUS*

Feed ingredient	Percentage digestibility			
	Protein	Fat	Carbohydrate	Gross energy
Fish meal	84.8	97.8	—	87.4
Fish meal plus corn	84.9	—	—	—
Meat and bone meal	77.7	—	—	68.7
Soybean meal	94.4	—	53.5	72.5
Corn (uncooked)	83.8	89.9	45.4	55.5
Corn (uncooked, mixed with fish meal)	—	—	65.4	—
Corn (cooked)	78.6	—	72.2	67.8
Wheat	89.6	84.9	60.8	65.3
Wheat bran	70.7	—	—	—
Alfalfa meal	65.7	—	27.7	22.9
Coffee pulp	29.2	—	—	11.4

Source: Popma 1982.

diet containing 9.4 kcal DE/g protein. Kubaryk (1980) reported that small *O. niloticus* (1.7–7 g) grew maximally when the DE/P ratio was 8.3 kcal/g for a 36% protein diet. He also found that as DE content of the diet increased, food consumption decreased, but that the amount of protein in the diet did not affect consumption rate.

Tilapia digest the gross energy in most commercial feedstuffs relatively well (Table 8.4). They do not digest highly fibrous feedstuffs, such as alfalfa meal and coffee pulp, well for energy needs. They digest carbohydrates in feedstuffs relatively well, much better than salmonids, but as the percentage of starch in the diet increases, digestibility decreases. Fats or proteins are more digestible to tilapias than are carbohydrates.

Essential fatty acids.   Tilapias appear to have a dietary requirement for fatty acids of the linoleic (n-6) family. Supplementation of tilapia diets with vegetable oils (soybean or corn oils) rich in 18:2 n-6 has given better performance than those containing fish oils high in 20:5 n-3 fatty acids (Takeuchi, Satoh, and Watanabe 1983a). The optimum dietary levels of n-6 fatty acids have been estimated to be about 1% for *T. zillii* (Kanazawa et al. 1980) and 0.5% for *O. niloticus* (Takeuchi, Satoh, and Watanabe 1983b). Deficiency signs observed in fish fed diets deficient in n-6 and n-3 fatty acids were poor appetite; retarded growth; and swollen, pale, and fatty livers. Tilapias do not tolerate as high a level of dietary fat as do salmonids. A dietary lipid level in excess of 12% depressed growth of juvenile *O. aureus* × *O. niloticus* hybrids (Jauncey and Ross 1982).

Vitamins.   Relatively little information is available on vitamin requirements of tilapias. One reason for this is that most tilapia culture is in ponds where the fish consume large quantities of natural foods, which probably satisfies their vitamin needs. Vitamin supplements are often deleted from practical feeds for tilapias cultured under extensive conditions in ponds. In intensive culture systems, where limited or no natural food organisms are present, supplemental vitamins must be added to commercial feeds. Due to the lack of information on vitamin requirements for tilapias, allowances established for other warmwater species are used.

Metabolically tilapias probably have similar vitamin requirements to other warmwater species. They show the classical vitamin C deficiency signs when deprived of the vitamin in the absence of natural foods. Vitamin E deficiency causes reduced growth rate, ceroids in liver and kidney, failure of mature males to develop sexual coloration, and degenerative changes in skeletal muscle. Lovell and Limsuwan (1982) showed that *O. niloticus* produced vitamin B_{12} in their long intestinal tract through bacterial synthesis and did not require the vitamin in their diet. Other vitamins may possibly be synthesized by intestinal microorganisms.

In view of the scarcity of information on vitamin requirements of tilapias, the vitamin allowances presented in Table 8.5 are recommended for tilapia diets that are fed in experimental or commercial cultures where natural food is absent or limited. These vitamin

Table 8.5. VITAMIN ALLOWANCES FOR TILAPIA DIETS

Vitamin	Amount per kilogram dry diet	
	Commercial feed (per kg)	Semipurified diets (per kg)
Vitamin A	4,000 IU	4,000 IU
Vitamin D	2,000 IU	2,000 IU
Vitamin E	50 IU	50 IU
Vitamin K	10 mg	10 mg
Choline	500 mg	3,000 mg
Niacin	30 mg	30 mg
Riboflavin	15 mg	15 mg
Pyridoxine	10 mg	10 mg
Thiamin	10 mg	10 mg
Pantothenic acid	50 mg	50 mg
Folacin	5 mg	5 mg
Vitamin C	200 mg	200 mg
Biotin	0	1 mg
Vitamin B_{12}	0	20 ug
Inositol	0	400 mg

Note: These allowances are based on requirements determined for other warmwater species, except for vitamins E, C, and B_{12}.

quantities, which allow for processing and storage losses, are based primarily on requirements for other warmwater species and have been used successfully in commercial and laboratory diets for tilapias.

Minerals.   Tilapias probably require the same minerals as other fish species for tissue formation, metabolism, and to maintain osmotic balance between the body fluid and the water. Like other finfishes, they probably get a significant amount of certain minerals, such as calcium, from the water.

Although there is a lack of information on the mineral requirements for tilapias, it is likely that they require all or most of the minerals known to be essential for other finfish species (chapter 2). In the absence of information on the mineral requirements for tilapia, the mineral requirements for channel catfish (Table 2.8 in chapter 2, page 64) can be used with reasonable assurance that mineral needs are met. The dietary level of available phosphorus required for maximum growth and normal bone mineralization of *O. niloticus* was estimated to be less than 0.9% (Watanabe et al. 1980). They also reported that the digestibility phosphorus in white fish meal for this species was about 65%.

FEEDS AND FEEDING

Natural foods.   Fish of the genus *Tilapia* are macrophyte-feeders, in which the adults feed mainly on filamentous algae and higher aquatic plants (Figure 8.4). Tilapias of the genus *Oreochromis* are microphagous; their feeding regime consists notably of phytoplankton, zooplankton, detritus, and benthic organisms. Species of this genus, such as *O. aureus, O. niloticus,* and *O. mossambicus,* are primarily omnivores. However, there is a great deal of overlapping among the diet compositions of various species of tilapias. For example, tilapias that feed on macrophytes also ingest the attached algae, bacteria, and detritus. Epiphyte consumers also frequently ingest the supporting macrophytes. Bacteria and protozoans, attached to detrital particles, are important sources of nutrients for benthic feeders. Animal components such as zooplankton and benthic organisms may also be eaten. Proximate composition of some aquatic organisms is given in Table 8.6. The digestibility values of crude protein of various natural foods by tilapias are presented in Table 8.7. Some algae are relatively high in protein and energy with good digestibility. The animal organisms are very good sources of protein and lipid with high digestibility.

Fig. 8.4. Tilapias consume a variety of foods. This *Tilapia rendalli* has been eating leaves from an alacosia plant. Partially eaten leaves can be seen in the pond in the background.

Most tilapias have short and widely spaced gill rakers, but are efficient in ingesting phytoplankton, even *Nannochloris,* a solitary coccoid green alga measuring less than 5 μ in diameter. The collecting processes of the minute food particles involve entrapment of algae in mucus secreted by mucous glands in the mouth and/or by filtration by microbranchiospines present on the outermost gill arches (Fryer and Iles 1972).

Most tilapia culture in the world is in ponds fertilized with organic or inorganic fertilizers. Under these conditions, natural food organisms supply substantial amounts of nutrients required by fish. Schroeder (1983) used stable carbon isotope analyses of the fish and the food sources and found that natural foods contributed 50% to 70% of the growth of tilapia polycultured with carps in ponds receiving manures and supplemental feeds. Stomach analyses show that up to 50% of stomach contents of tilapia is natural food in intensively fed pond cultures, indicating that the natural pond productivity contributes substantial amounts of nutrients for the growth of tilapias.

Table 8.6. COMPOSITION OF AQUATIC ORGANISMS CONSUMED BY FISH

Organism	Moisture	Percentage of dry matter			
		Protein[1]	Lipid	Crude fiber or chitin[2]	Ash
Plants					
Aphanizomeno flos-aquae	94	48	14	8	7
Spirogyra sp.	94	13	10	22	24
Ceratophylum demersum	90	9	5	11	15
Animals					
Daphnia puplex	94	49	16	6	19
Diaptomus sp.	92	58	24	6	5
Chirocephalopsis bundyi	93	59	17	3	10
Gammarus lacustris	85	39	10	7	27
Hyalella azteca	84	36	9	7	30
Chaoborus americanus (nymphs)	93	60	17	6	7
Aeshna sp. (nymphs)	86	34	21	7	4
Enallagma boreale (nymphs)	86	58	13	10	5
Enallagma cyathigerum (nymphs)	86	58	13	9	5
Siraga sp.	74	55	32	10	3
Chironomid larvae	86	47	13	4	9

[1] Does not include chitin nitrogen.
[2] Crude fiber in plants and chitin in animals.
Source: Yurkowski and Tabachek (1979).

Table 8.7. DIGESTIBILITY OF CRUDE PROTEIN IN
PHYTOPLANKTON BY *S. AUREUS*

Phytoplankton	Percentage digestibility
Filamentous green algae:	
Hydrodictyon sp.	70
Oedogonium sp.	63
Planktonic green algae:	
Dictyosphaerium sp.	43–52
Sphaerocystis sp.	39–43
Scenedesmus sp.	42–64
Nannochloris sp.	32
Volvox sp.	68
Planktonic blue-green algae:	
Microcystis sp.	57–68

Sources: From Manandhar (1977) and Popma (1982).

Practical feeds.   Commercial pond feeds for tilapias usually contain 24% to 28% protein. Natural pond foods contribute a significant amount of protein, so this level is assumed to be high enough. However, few pond studies have been conducted to compare various diet formulations for extensive or semi-intensive culture of tilapias. A 25% protein pellet composed of 15% fish meal, 20% soybean meal, 20% ground wheat, and 45% ground sorghum has been used successfully in Israel in this type of production. The importance of micronutrient supplementation in pond feeds for tilapia is not well known. Due to the extreme variation in the culture practices used, formulation of practical feeds to efficiently supplement the nutrient contribution of the natural food is practically impossible. In intensive culture, such as in raceways or cages, tilapias rely solely on the prepared feeds as a source of nutrients. Thus, a nutritionally complete feed containing all essential nutrients is required. Model formulae for pond and raceway feeds for tilapia are given in Table 8.8. Commercial diets formulated for common carp and channel catfish have been fed successfully to tilapia.

Tilapia accept a variety of feeds, in meal form and in moist, sinking, and floating pellets. Crude feedstuffs, such as rice bran, are offered in meal form, whereas the compounded diets are most often processed into pellets. Tilapia can utilize meal type feeds effectively, although they obviously do not eat all of the meal that is put into the water. Crude feed sources may be uneconomical when pelleted for pond feeding of tilapias; however, high quality feeds should be processed into pellets to minimize waste.

Table 8.8. MODEL FORMULAE FOR POND (25% PROTEIN)
AND RACEWAY (32% PROTEIN) FEEDS FOR TILAPIA

Ingredient	Pond feed, %	Raceway feed, %
Fish meal, anchovy, or menhaden	8.0	12.0
Soybean meal (48% protein)	28.8	43.0
Grains or grain byproducts	59.4	38.8
Fat	—	1.9
Pellet binder	2.0	2.0
Dicalcium phosphate	1.5	2.0
Vitamin mix[1]	0.25	0.25
Mineral mix[2]	0.05	0.05

[1] Vitamin mix should provide the vitamin allowances for a commercial tilapia feed given in Table 8.5.
[2] Mineral mix should provide the following minerals in the amounts in mg/kg of feed: manganese, 25; zinc, 100; iron, 44; copper, 3; iodine, 5; cobalt, 0.05; and selenium, 0.3.

The physical properties of pelleted tilapia feeds are important, especially water stability and size. The feeds must remain stable in the water for a sufficient period for the fish to consume and to minimize the losses of nutrients through dissolution and wastage of feed. Hard and durable pellets are necessary when the feeds are to be crumbled for feeding smaller fish. Tilapia seem to prefer smaller pellets than channel catfish and salmonids of comparable size. They tend to chew the pellets rather than swallow them whole as do most finfish species. Unacceptable pellets are usually taken into the mouth and then rejected several times before they are finally consumed or discarded. For feeding tilapias to marketable size of 500 g, the most common pellet size is approximately 3mm to 4mm in diameter and 6mm to 10mm in length. Feeds in meal or crumble forms are used for fry and fingerlings. These are made by first pelleting or extruding the feed mixture and then reducing the particles to a desirable size by crumbling.

Feeding rates for tilapia are affected by species, size, temperature, feeding frequency, and the availability of natural foods. *T. rendali* consume more feed than *O. niloticus* of comparable age (Balarin and Haller 1982). As with other fishes, feed consumption rate of tilapias is inversely related to fish size. Tilapia benefit from multiple daily

Table 8.9. FEEDING RATES AND FREQUENCIES FOR VARIOUS SIZES OF TILAPIAS AT 28° C

Size	Daily feeding (% of fish weight)	Times fed daily
2 days old to 1 g	30–10	8
1–5 g	10–6	6
5–20 g	6–4	4
20–100 g	4–3	3–4
>100 g	3	3

Sources: From Jauncey and Ross (1982), Coche (1982), and Kubaryk (1980).

feedings. Because of their continuous feeding behavior and smaller stomach capacity, tilapia respond to more frequent feeding than channel catfish and salmonids. Kubaryk (1980) found that *O. niloticus* grew faster when fed four times daily than when fed two times, but did not grow faster when fed eight times. Tilapia will consume more feed and grow faster than channel catfish or salmonids. Recommended feeding rates and frequencies for various sizes of tilapia are given in Table 8.9.

Feeds are offered to fish by hand, blower, demand feeders, or automatic feeders. Hand feeding is labor intensive, but has the advantage over other methods in that the feeder can observe the fish feeding better. Demand feeders are feeding devices that deliver feed when fish activate the feed release device. Automatic feeders deliver measured quantities of feed at various time intervals. Meriwether (1986) showed that *O. niloticus* gained 72% more weight when fed by demand feeder than when fed by hand one time daily, but feed conversion was 45% poorer for the fish fed by demand feeder. Demand feeders have disadvantages in that fish may cause release of the feed without consuming it.

REFERENCES

ALLISON, R., R. O. SMITHERMAN, and J. CABRERO. 1976. Effect of high density culture on reproduction and yield of *Tilapia aurea*. *FAO* Technical Conference on Aquaculture, Kyoto, Japan. FIR: AQ/Conf/76/E.47.

BALARIN, J. D., and R. D. HALLER. 1982. The intensive culture of tilapia in tanks, raceways and cages. In *Recent advances in aquaculture,* eds. J. E. Muir and R. J. Roberts, 265–356. Boulder, CO: Westview Press.

CHEN, T. P., and Y. LI. 1980. Integrated agriculture-aquaculture studies in Taiwan. In *Integrated agriculture-aquaculture farming systems,* eds. R. S. V. Pullin and Z. H. Shehadeh, 239–241. Manila, Philippines: ICLARM.

COCHE, A. G. 1982. Cage culture of tilapias. In *The biology and culture of Tilapia,* eds. R. S. V. Pullin and R. H. Lowe-McConnel. Manila, Philippines: ICLARM.

COLLIS, W. J., and R. O. SMITHERMAN. 1978. Production of tilapia hybrids with cattle manure or commercial diet. Proceedings of Symposium on Culture of Exotic Fishes, Atlanta, GA, pp. 43–54.

CRUZ, E. M., and Z. H. SHEHADEH. 1980. Preliminary results of integrated pig-fish and duck-fish production tests. In *Integrated agriculture-aquaculture farming systems,* eds. R. S. V. Pullin and Z. H. Shehadeh, 225–238. Manila, Philippines: ICLARM.

DELA CRUZ, C. R. 1980. Integrated agriculture-aquaculture farming systems in the Philippines, with two case studies on simultaneous and rotational rice-fish culture. In *Integrated agriculture-aquaculture farming systems,* eds. R. S. V. Pullin and Z. H. Shehadeh, 209–223. Manila, Philippines: ICLARM.

FRYER, G., and T. D. ILES. 1972. *The cichlid fishes of the Great Lakes of Africa.* Hong Kong: T.F.H. Publications, Inc. Ltd.

HALEVY, A. 1979. Observation on polyculture of fish under standard farm pond conditions at the Fish and Aquaculture Research Station, Dor, during the years 1972–1977. *Bamidgeh* 31(4): 96–104.

JACKSON, A. J., B. S. CAPPER, and A. J. MATTY. 1982. Evaluation of some plant proteins in complete diets for the tilapia *Sarotherodon mossambicus. Aquaculture* 27(2): 97–109.

JAUNCEY, K., and B. ROSS. 1982. *A guide to tilapia feed and feeding.* Institute of Aquaculture, University of Sterling, Scotland.

KANAZAWA, A., S. I. TESHIMA, M. SAKAMOTO, and MD. A. AWAL. 1980. Requirements of *Tilapia zillii* for essential fatty acids. *Bull. Jap. Soc. Sci. Fish.* 46(11): 1353–1356.

KUBARYK, J. M. 1980. Effect of diet, feeding schedule and sex on food consumption, growth and retention of protein and energy by tilapia. Ph.D. Diss., Auburn University, Auburn, AL.

LIAO, I-CHIU, and T. P. CHEN. 1983. Status and prospects of tilapia culture in Taiwan. *Proceedings of the International Symposium on Tilapia in Aquaculture,* Nazareth, Israel, 8–13 May 1983: 588–598.

LOVELL, R. T., and T. LIMSUWAN. 1982. Intestinal synthesis and dietary nonessentiality of vitamin B_{12} for *Tilapia nilotica. Trans. Am. Fish. Soc.* 11: 485–490.

LOVSHIN, L. L. 1982. Tilapia hybridization. In *The biology and culture of tilapias,* eds. R. S. V. Pullin and R. H. Lowe-McConnell, 279–307. Manila, Philippines: ICLARM.

MANANDHAR, H. N. 1977. Digestibility of phytoplankton by silver carp and three tilapias in polyculture with channel catfish. Master's Thesis, Auburn University, Auburn, AL.

MERIWETHER, F. H. 1986. An inexpensive demand feeder for cage-reared tilapia. *Prog. Fish. Cult.* 48: 226.

POPMA, T. J.   1982.   Digestibility of selected feedstuffs and naturally occurring algae by tilapia. Ph.D. Diss., Auburn University, Auburn, AL.

POPMA, T. J., F. E. ROSS, B. L. NERRIE, and J. R. BOWMAN.   1984.   *The development of commercial farming of tilapia in Jamaica, 1979–1983*. Research and Development Series No. 31, International Center for Aquaculture, Auburn University, Auburn, AL.

SCHROEDER, G. L.   1983.   The role of natural foods in tilapia growth: A study based on stable isotope analyses. *Proceedings of the International Symposium on Tilapia in Aquaculture*, Nazareth, Israel, 8–13 May 1983: 313–322.

SIN, A. W. C., and M. T. CHIU.   1983.   The intensive monoculture of the tilapia hybrid, *Sarotherodon nilotica* (male) × *S. mossambica* (female) in Hong Kong. *Proceedings of the International Symposium on Tilapia in Aquaculture*, Nazareth, Israel, 8–13 May 1983: 506–516.

TAKEUCHI, T., S. SATOH, and W. WATANABE.   1983a.   Dietary lipids suitable for practical feed of *Tilapia nilotica*. *Bull. Jap. Soc. Sci. Fish*. 49(9): 1361–1365.

TAKEUCHI, T., S. SATOH, and W. WATANABE.   1983b.   Requirement of *Tilapia nilotica* for essential fatty acids. *Bull. Jap. Soc. Sci. Fish*. 49(7): 1127–1134.

TAL, S., and I. ZIV.   1978.   Culture of exotic fishes in Israel. *Symposium on Culture of Exotic Fishes*, Atlanta, GA, January 4, 1978: 1–9.

TREWAVAS, E.   1982.   Tilapias: Taxonomy and speciation. In *The biology and culture of tilapias*, eds. R. S. V. Pullin and R. H. Lowe-McConnel, 3–14. Manila, Philippines: ICLARM.

VIOLA, S., and Y. ARIELI.   1983.   Nutrition studies with tilapia (*Sarotherodon*). 1—Replacement of fish meal by soybean meal in feeds for intensive tilapia culture. *Bamidgeh* 35(1): 9–17.

WATANABE, T., T. TAKEUCHI, A. MURAKIMI, and C. OGINO.   1980.   The availability to *Tilapia nilotica* of phosphorus in white fish meal. *Bull. Jap. Soc. Sci. Fish*. 46(7): 897–899.

WINFREE, R. A., and R. R. STICKNEY.   1981.   Effect of dietary protein and energy on growth, feed conversion efficiency and body composition of *Tilapia aurea*. *J. Nutr*. 111(6): 1001–1012.

WU, J. L., and L. JAN.   1977.   Comparison of the nutritive value of dietary proteins in *Tilapia aurea*. *J. Fish Soc. Taiwan* 5(2): 55–60.

YURKOWSKI, M., and J. L. TABACHEK.   1979.   Proximate and amino acid composition of some natural fish foods. In *Finfish nutrition and fishfeed technology*, Vol. I, eds. J. H. Halver and K. Tiews, 435–448. Berlin: Heenemann Verlagsgesellschaft mbH.

9

Practical Feeding—Salmon and Trout

Ronald W. Hardy

Salmonid aquaculture is the oldest form of fish rearing in North America, with recorded efforts at artificial propagation of salmon and trout dating back to the 1850s in Canada and the 1870s in California. For many years, the production level and dollar value of salmonid aquaculture were greater than all other forms of aquaculture. Currently, salmonid aquaculture in North America is second in production behind catfish and third in dollar value behind catfish and ornamental and bait fishes. The salmonids include the five species of Pacific salmon, Atlantic salmon, trouts, and chars.

TYPES OF SALMONID CULTURE

Of the many species of trout and char found in North America (Table 9.1), the rainbow trout is by far the most extensively cultured species, both for stocking of public waters for recreational fishing and for food production. Over 300 state and federal hatcheries rear trout for fisheries enhancement. Production of trout for stocking is about 200 million fish annually. Production of food fish is approximately 45–50 million pounds per year. Production is spread throughout many states in the northern and central United States; however, over 80% of commercial trout culture is located in the Hagarman Valley of Idaho (see Figure 9.1). The reason for this location is the availability of enormous amounts of constant-temperature, fully oxygenated, ground water along the Snake River.

Pacific salmon, in contrast to trout, are primarily cultured for public fisheries enhancement. Pacific states rear and release about 600 million salmon fingerlings, or smolts, each year. British Columbia, Canada, contributes an additional 400 million. The different species of

Table 9.1. TROUT AND CHAR RAISED IN HATCHERIES IN NORTH AMERICA

Common name	Scientific name	Comments
Rainbow trout	*Salmo gairdneri*	Raised extensively for food and for planting
Brown trout	*Salmo trutti*	Raised mainly for planting
Cutthroat trout	*Salmo clarki*	Raised mainly for planting
Golden trout	*Salmo aquabonita*	Limited culture
Atlantic salmon[1]	*Salmo solar*	Raised for food and for planting
Brook trout	*Salvelinus fortunalis*	Raised for planting only
Lake trout	*Salvelinus namaycush*	Raised for planting only
Dolly Varden	*Salvelinus malma*	Raised for planting only
Sunapee trout	*Salvelinus aureolus*	Raised for planting only

[1] The Atlantic salmon is classified by taxonomists as a trout.

Pacific salmon are reared in freshwater hatcheries for different lengths of time, depending upon their natural periods of freshwater residence (Table 9.2). During late spring and summer, as water temperature and photoperiod increase, the salmon fry and fingerlings undergo a subtle

Fig. 9.1. Commercial trout farm in the Hagarman Valley of Idaho. Rainbow trout are grown to marketable size as food fish in raceways furnished with a continuous flow of oxygenated ground water.

Table 9.2. PACIFIC SALMON INDIGENOUS TO NORTH AMERICA

Common name	Scientific name	Size at release (g)	Size as adults (kg)	Age at return (yr)	Percentage of commercial catch	Percentage of hatchery contribution to commercial catch
Chinook salmon	Oncorhynchus tshawytscha	5–10	1–55	3–5	2.5	40
Coho salmon	Oncorhynchus kisutch	25–40	3–6	3–4	8.3	46
Chum salmon	Oncorhynchus keta	0.5–2	1–20	3–4	10.1	5
Pink salmon	Oncorhynchus gorbuscha	0.5–2	1–4	2	52.6	Unknown
Sockeye salmon	Oncorhynchus nerka	5–25	2–4	2–4	26.5	Unknown

metamorphosis that results in a transformation from the freshwater to the saltwater-tolerant form (smolts). The fish are then released into streams and rivers and voluntarily migrate to the ocean. After a period of near-shore feeding, they move offshore into the North Pacific, where they remain for several years until they return to near-shore areas during the summer and fall as mature adults preparing to migrate back to freshwater to spawn and die. It is at this point that the adult salmon enter the commercial and sport fisheries. Hatchery-raised fish make a significant contribution to the fishery, as freshwater habitat loss has reduced wild populations.

Commercial culture of Pacific salmon in the United States is currently in its infancy and is primarily concentrated on the produc-

Fig. 9.2. Atlantic salmon (7 kg) raised in net pen in seawater.

tion of pan-sized coho salmon. The coho salmon is the least difficult of the Pacific salmon to raise in marine net-pens. Improvements in aquaculture technology, genetic selection, feeds, disease prevention and treatment, and marketing will likely result in a further expansion of commercial salmon culture from its current level of about 1 million pounds per year.

In Europe, commercial marine net-pen culture of Atlantic salmon and rainbow trout is a successful industry (see Figure 9.2). Norway produces over 75% of the farm-raised salmon grown worldwide, with a production of over 40,000 metric tons in 1987. In contrast to the North American production of primarily pan-sized fish, European salmonid culture produces larger fish, from 1 kg to 10 kg, with the most common harvest size of 3 kg to 5 kg.

PRODUCTION OF SEEDSTOCK

Salmon and trout have similar life histories and are relatively easy to culture compared to many fish. At maturity, the gonad weight of salmonids is about 20% to 25% of the total weight of the fish. Eggs become loose in the peritoneal cavity and can be expelled by applying pressure on the upper ventral surface of the fish. Trout can survive spawning, while salmon die after spawning. Trout eggs are removed from the fish by pressure or flushing while salmon are normally killed and the eggs are removed by cutting open the fish. The eggs are placed in a container and sperm is expelled onto them by squeezing the male. Water is added to water-harden the eggs, and the fertilized eggs are placed in wire baskets or trays in upwelling fresh water to incubate. After 1 day, unfertilized eggs turn white and can be removed. The remaining eggs are allowed to develop undisturbed until they reach eyed stage (when the eye pigments can be easily seen), at which time they can be moved. Trout eggs hatch in about 75 days at 5° C, while at 10° C, hatching occurs in about 31 days. About 20–30 days after hatching, the yolk-sac fry (alevins) have absorbed most of the egg yolk and the fish become free-swimming and begin to seek food on the surface of the water.

First feeding (swim-up) fry range in size from about 150 mg for trout to 300 mg to 500 mg for salmon fry. Swim-up fry are placed in shallow tanks and fed a mash type feed. As the fish grow, the feed particle size is increased. Frequent feeding in small fish is of critical importance; initially, the small fry are fed almost continuously. As the fish grow, feeding frequency is reduced. Once all fry are actively feeding and growing, they are placed in larger ponds, raceways, or tanks supplied with constantly running fresh water. Salmon destined for release are fed so that they reach a target size by their hatchery release date. This may involve adjusting the feeding rate to insure that the fish do not get too large.

GENERAL CULTURE METHODS

Salmon and trout may be raised in either freshwater ponds or in floating net-pens in saltwater, although salmon grow more rapidly in saltwater (see Figure 9.3). Coho and chinook salmon smolts can tolerate abrupt saltwater transfer, although some losses occur. Atlantic salmon and trout must be acclimated to saltwater by gradual exposure to increasing strength saltwater. After saltwater transfer, the fish are fed either dry or moist diets, depending in part on the species of fish. Chinook salmon grow faster in saltwater when they are fed moist diet formulations, while coho salmon, Atlantic salmon, and rainbow trout do well on either moist or dry diets.

Salmon and trout reared as broodstock are fed at a high rate to insure rapid growth and early maturity, which is somewhat size-related in salmon and trout. Coho salmon raised in captivity mature as 2-year-old fish, while chinook mature at 3 years to 4 years. Atlantic salmon mature as 4- to 5-year-old fish. Rainbow trout mature at age 2 if the growth rate is rapid, but more often maturation is delayed until age 3. As the fish approach the onset of maturation, somatic growth stops and rapid gonadal growth begins. This switch occurs about 6 months to 8 months before spawning, and it is at this time broodstock fish should be switched from a production-type diet to a broodstock diet.

NUTRIENT REQUIREMENTS

Most the research on the nutrient requirements of salmon and trout has been conducted on small fish raised in freshwater under laboratory conditions. In general, nutrient requirements are determined by feeding nonstressed small fish semi-purified diets that contain all of the known required nutrients except the one being tested. The dietary level of the nutrient being tested is varied, and fish growth or some other physiological parameter is measured. The dietary level at which additional supplementation does not result in increased health or performance is considered to be the minimum dietary requirement.

Fig. 9.3. Commercial marine net-pen salmon culture operation in Puget Sound on the coast of Washington.

Protein and amino acid requirements. Dietary protein requirement for salmonids varies with age, dietary energy level, and balance of amino acids. Theoretically, salmon and trout do not require dietary protein per se, but rather, require only the indispensable amino acids contained in protein. In practical feeds, the minimum dietary protein level for good growth is usually around 45% to 50% for swim-up fry, 40% to 45% for larger fry, and 35% to 40% for fingerlings up to harvest or planting size (Hilton and Slinger 1981).

Salmon and trout require 10 amino acids in their diet. The other amino acids needed for protein synthesis by the fish can be synthesized from intermediates of metabolism as long as sufficient amine groups are available. Essential amino acid requirements of chinook salmon, determined by the traditional method of feeding purified amino acid test diets and measuring growth response, are presented in Table 2.4 in chapter 2 (page 26). Ogino (1980) estimated amino acid requirements of rainbow trout by measuring daily increase in essential amino acids in the body of fish fed high quality proteins. The values for rainbow trout are compared with those for chinook salmon (as determined from growth responses) in Table 9.3. The data in the table indicate marked differences in requirements for certain amino acids, especially the sulfur amino acids and arginine.

Energy.   Because energy intake influences amount of food consumed, the energy content of the diet and the sources of dietary energy merit careful consideration. Salmon and trout use dietary proteins and lipids for energy well, but assimilate carbohydrates poorly. Lipids are generally a less expensive source of energy for salmonids than are protein and carbohydrate sources, when compared on the basis of metabolizable energy (ME) provided. In a 35% to 40% protein diet, balanced in essential amino acids, the optimum ratio of protein to energy for trout fingerlings is provided with 15% to 20% lipid (Watanabe, Takeuchi, and Ogino 1979).

Essential fatty acids.   Salmon and trout require about 1% to 2% omega-3 fatty acids in the diet to prevent essential fatty acid deficiency signs (Castell 1979; Watanabe 1982). These include poor growth, pale liver, depigmentation, caudal fin necrosis, "shock syndrome" in which fish become unconscious during stress, and elevated tissue levels of omega-9 (n-9) fatty acids. These signs can be prevented by any of the n-3 fatty acids, although growth rates are reportedly higher when

Table 9.3. QUANTITATIVE AMINO ACID REQUIREMENTS OF CHINOOK SALMON AND RAINBOW TROUT

	Salmon		Trout	
Amino acid	Percentage of the protein	Percentage of the diet (40% protein)	Percentage of the protein	Percentage of the diet (40% protein)
Arginine	6.0	2.4	3.5	1.4
Histidine	1.8	0.7	1.6	0.6
Isoleucine	2.2	0.9	2.4	1.0
Leucine	3.9	1.6	4.4	1.8
Lysine	5.0	2.0	5.3	2.1
Methionine	4.0[1]	1.6[1]	1.8[3]	0.7[3]
Phenylalanine	5.1[2]	2.1[2]	3.1[4]	1.2[4]
Threonine	2.2	0.9	3.4	1.4
Tryptophan	0.5	0.2	0.5	0.2
Valine	3.2	1.3	3.1	1.2

[1] In the absence of cystine, which can substitute for about 1/3 of the methionine requirement.
[2] In the absence of tyrosine, which can substitute for a portion of the phenylalanine requirement.
[3] In the presence of 0.4% cystine.
[4] In the presence of 0.8% tyrosine.
Source: Salmon requirements from NRC (1981). Trout requirements from Ogino (1980).

long-chained highly unsaturated fatty acids of the n-3 group, such as C20:5, C22:6, are fed. Essential fatty acid deficiency signs can be exacerbated by high dietary levels of n-6 fatty acids. Usually, a level of 4% to 5% marine fish oil in the diet provides a sufficient dietary level of n-3 fatty acids for salmonids.

Vitamin requirements. Salmon and trout require 15 vitamins in their diet to insure good growth and optimal health. The precise quantitative dietary vitamin requirements for salmon and trout of various sizes and reared in various environments have not been established, but vitamin requirements of small fish fed semipurified diets in a laboratory environment have been determined (Table 2.6, chapter 2, page 33). Characteristic vitamin deficiency signs determined with small salmonids in a controlled environment have been described (Table 2.5, chapter 2, page 30). Supplementing practical diets to insure these levels after pelleting and storage has proven effective in preventing overt vitamin deficiency signs in practical fish culture.

Mineral requirements. For many years, it was thought that salmonids obtained a large proportion of their mineral needs directly from the water in which they lived. In the case of some minerals that are present in high concentration in dissolved form, like calcium, the water can make a significant contribution to the requirement, but for most minerals the water is not a major source and they must be present in the diet of salmonids to prevent mineral deficiencies (Lall 1979). Most practical feeds that contain a significant amount of fish meal have sufficient levels of essential elements to satisfy the requirements of growing salmon and trout. Some dietary components, such as phytic acid from plant seeds and high-ash fish meal, can reduce the availability of certain divalent cations, such as zinc, to the fish and cause deficiency problems. For example, diets containing high levels of fish meal or soybean meal should be overfortified with zinc. The known mineral requirements of salmonids and characteristic deficiency signs are shown in Table 9.4.

FEED FORMULATION

Several important developments in salmonid nutrition and diet formulation have occurred that have enabled salmonid aquaculture to advance to current levels of production and sophistication. They are (a) development of a semi-purified test diet for salmonids that allowed the

Table 9.4. DIETARY MINERAL REQUIREMENTS OF SALMONIDS

Mineral	Dietary requirement	Deficiency signs
Calcium	0.3%	None described
Phosphorus	0.5–1.0%	Poor growth and bone development; low bone phosphorus
Magnesium	0.05–0.07%	Poor growth; low tissue magnesium, especially bones; spinal deformity
Iron	25–50 mg/kg	Hypochromic microcytic anemia
Manganese	12–13 "	Short body dwarfism; depressed growth
Copper	1–4 "	None described
Selenium	0.03–0.1 "	High fry mortality; reduced serum glutathionine peroxidase activity
Iodine	0.1–0.3 "	Thyroid hyperplasia
Zinc	20–100 "	Lens cataracts; poor growth; low tissue zinc concentration

quantitative nutrient requirements to be determined, (b) development of pelleted dry feeds for salmonids, and (c) development of the Oregon Moist Pellet. These developments have resulted in economical, readily available feeds that support good growth and prevent nutritional deficiencies.

Early salmonid feeds.   Prior to the development of dry pelleted feeds, salmonids were fed wet feeds made from beef liver or other slaughterhouse byproducts, fish or fish products, and other materials that were available to the hatchery. Part of the work of the hatchery staff was to make fish food. Formulas such as 48% ground scrap fish, 28% liver, and 24% salmon eggs, or 47.5% fresh liver, 47.5% canned carp, and 5% dried brewer's yeast were used. The feed was formed by hand into chunks and fed fresh. As the demand for ingredients outstripped supply, later formulations used dry feed mixtures in combinations with chopped fish and liver. Disease transmission via the feed was an ever-present threat when formulations included raw fish.

Moist feeds.   Development of the Oregon Moist Pellet (OMP) was a natural evolutionary step in production of soft, acceptable diets for young salmon made from wet-mix/dry-mix combinations, but it was an important advancement. A major improvement over earlier feeds was the hydrolysis and pasteurization of the fish processing waste or trash fish in the wet mix. Another was the desirable physical properties of the resulting semi-moist mixtures which could be extruded into pellets and frozen, sacked, and shipped to hatcheries from a feed plant. These developments eliminated the transmission of disease in the feed and eliminated the need for each hatchery to prepare its own feed. The OMP formula has been modified since its development in the late 1950s due to changes in availability and price of ingredients; a current formula is presented in Table 9.5. Despite the fact that the OMP is higher in price than dry salmonid feeds, many hatcheries continue to use it because of its high acceptability by young salmon, especially at low water temperatures. Several feed manufacturers are now producing semi-moist feeds (12–25% moisture) that contain mold inhibitors and preservatives that permit storage without freezing.

The commercial Atlantic salmon pen-culture industries of Europe use dry and semi-moist feeds. The dry feeds are high-protein, high-fat pelleted formulations. The semi-moist feeds are made from chopped, frozen capelin mixed with a dry meal containing other feed ingredients, binders, vitamins, and mineral premixes. The semi-moist feeds can be mixed and extruded into pellets and fed fresh at the salmon farm. Substituting a part or all of the chopped fish with liquefied fish products, such as acid-preserved, hydrolyzed fish processing waste (fish silage), is an increasing practice. The ratio of wet material to dry mix is about 60:40.

Table 9.5. OREGON MOIST PELLET DIET FORMULATIONS

Ingredient	Oregon mash OM-3 (%)	Oregon pellet OP-4 (%)	Oregon pellet OP-2 (%)
Herring meal	49.9	Minimum 47.5	Minimum 14
Other fish meal	—	—	Minimum 14
Wheat germ meal	10.0	Remainder	Remainder
Dried whey	8.0	4.0	5.0
Cottonseed meal	—	—	Maximum 10
Poultry by-product meal	—	—	Maximum 8)
Corn distillers dried solubles	—	—	4.0
Sodium bentonite	—	3.0	—
Vitamin premix	1.5	1.5	1.5
Mineral premix	0.1	0.1	0.1
Wet mix [1]	20.0	30.0	30.0
Fish oil	10.0	6.5–7.0	6.0–6.7
Choline chloride	0.5	0.5	0.5
Proximate composition:			
Crude protein		>35	>35
Fat		>10	>10
Moisture		<35	<35
Crude fiber		< 4	< 4

Note: Diet must be stored frozen.
[1] Ground, heated fish, or fish processing waste.

Pigment enhancement. In commercial salmon culture, highly pigmented flesh is critical for market acceptability. A source of the carotenoid pigment astaxanthin, which imparts the characteristic pink color to the flesh, must be included in the diet. Natural sources of astaxanthin are processing wastes from shrimps, crabs, krill, and crawfish. These products can be added to the diet in moist or dry meal forms, or as the extracted carotenoids. Astaxanthin is also synthesized commercially. The synthetic carotenoid, canthaxanthin, can also be added to salmonid feeds to produce pigmented flesh. Canthaxanthin is slightly less red than astaxanthin. These pigment sources are added to salmonid feeds at levels to provide 40 mg to 60 mg of carotenoid/kg of feed. The carotenoid-containing feed is fed for 3 to 6 months to produce an acceptable salmon colored product.

Dry pellet feeds. Nearly all trout feeds, and a significant portion of salmon feeds, are made by compression steam-pelleting, although some trout feed is made by extrusion to form floating pellets. With the determination of the dietary vitamin requirements of salmonids, the last hurdle was cleared to permit the feeding of nutritionally complete dry feeds and eliminate the need to supplement with liver or other nutrient enriched products. Dry salmonid formulations contain fairly high levels of fish meal and are therefore more expensive than feeds made for domestic animals. Model trout and salmon feed formulations are presented in Tables 9.6 and 9.7.

Ingredients used in salmonid feeds. Feed formulations high in protein and fat generally limit the use of many common feed ingredients used in the feeds of domestic land animals. Ingredients that contain relatively high amounts of crude fiber or starch, or which are low in protein and metabolizable energy, cannot be included in salmonid feeds because to do so would limit the levels of other ingredients, thereby creating a feed with a low nutrient density. Thus, ingredients such as corn, wheat, or dehydrated alfalfa, are rarely used. In practice, salmonid diet formulations are usually limited to fish and meat meals, oilseed meals, grain milling byproducts, and low levels of other protein supplements, such as blood meal, feather meal, poultry byproduct meal, and brewer's yeast.

Use of some ingredients is restricted because of the presence of antinutritional factors, toxicants, contaminants, and palatability properties of some ingredients. Soybean meal, cottonseed meal, and canola meal are examples of ingredients that are often restricted to levels lower than theoretical levels calculated by least-cost formu-

Table 9.6. ABERNATHY DRY SALMON DIET FORMULATIONS

Ingredient	Starter diet, (S8.2) (%)	A18-2 (%)	A19-2 (%)
Fish meal (herring or anchovy)	58.0	55.0	50.0
Dried whey product	10.0	10.0	10.0
Wheat middlings	Remainder	Remainder	Remainder
Wheat germ meal	—	5.0	5.0
Spray-dried blood meal	10.0	10.0	10.0
Condensed fish solubles	3.0	3.0	3.0
Poultry byproduct meal	1.5	1.5	1.5
Vitamin premix[1]	1.5	1.5	1.5
Trace mineral premix[2]	0.05	0.1	0.1
Choline chloride	0.58	0.58	0.58
Ascorbic acid	0.1	0.1	0.1
Lignin sulfonate pellet binder	—	2.0	2.0
Fish oil	12.0	9.0	9.0
Proximate composition:			
Crude protein	>48	>45	>41
Crude fat	17–19	13–17	13–17
Moisture	<10	<10	<10

[1] Vitamin premix no. 2, containing the following in mg or IU/kg premix: d-Biotin, 39.6; vitamin B_{12}, 3.96; vitamin E, 33,400 IU; folacin, 847; myo-inositol, 8,800; vitamin K, 611.6; niacin, 14,674; d-pantothenic acid, 7,040; vitamin B_6, 2,057; riboflavin, 3,520; thiamin, 2,860; vitamin D_3, 29,337 IU; vitamin A, 440,000 IU.
[2] Trace mineral mix no. 3, containing the following in g/kg premix: $ZnSO_4$, 184.8; $MnSO_4$, 55; $CuSO_4$, 3.89; KIO, 16.9.

Table 9.7. U.S. FISH AND WILDLIFE SERVICE TROUT DIET FORMULATIONS

Ingredient	Starter diet SD9-30 (%)	Grower diet GR6-30 (%)	Grower diet GR7-30 (%)
Fish meal (herring or anchovy)	Minimum 50	Minimum 30	Minimum 20
Wheat flour[1]	10.0	5.0	5.0
Wheat middlings	—	17.5	37.5
Soybean flour or meal	15.0	25.0	15.0
Spray-dried blood meal	10.0	10.0	10.0
Trace mineral premix[2]	0.05	0.1	0.1
Vitamin premix[3]	0.6	0.4	0.4
Choline chloride	0.225	0.175	0.175
Ascorbic acid	0.075	0.075	0.075
Fish oil	12.0	10.0	10.0
Lignin sulphonate pellet binder	2.0	2.0	2.0
Proximate composition:			
Crude protein	>50	>42	>34
Crude fat	>17	>13	>13
Moisture	<10	<10	<10

[1] Total is brought to 100% by adjusting the level of wheat feed flour or wheat middlings.
[2] Same as used in Table 9.6.
[3] Vitamin premix no. 30, containing the following per kg premix: D-calcium pantothenate, 26.5 g/pyridoxine, 7.7 g; riboflavin, 13.2 g; niacinamide, 55.1 g; folic acid, 212 g; thiamin, 8.8 g; biotin, 88.2 mg; vitamin B_{12}, 55 mg; menadione sodium bisulfite, 2.76 g; vitamin E, 88,200 IU; vitamin D_3, 110,250 IU; vitamin A, 1,653,750 USP.

lation. Young salmon find diets containing even low levels of soybean meal unpalatable, thus soybean meal is not used in the diets of young salmon. Older salmon, however, will consume diets containing up to 20% soybean meal. Cottonseed meal can be fed at levels up to 15% in salmon and trout diets; higher levels are not recommended due to the free-gossypol content. Cottonseed meal is not recommended in the diets of broodfish. Canola meal contains compounds that impair thyroid function, so salmon diets should contain no more than 15% and trout diets should be restricted to 25%. Other ingredient restrictions include the use of high-ash (>15%) fish meal, although the deleterious effects of high-ash fish meal can be overcome by extra trace element fortification of the diet.

Broodstock feeds.   Very little is known about the nutritional requirements of broodstock salmonids, so broodstock diets are formulated to contain slightly higher levels of protein and energy than production diets. In addition, the diets are fortified with extra vitamins and trace elements, and carotenoid pigments are added to the formulation to enhance egg pigmentation. The need for carotenoid supplementation has not been firmly established, but a number of reports indicate a variety of potential roles of carotenoid pigments in reproduction (Meyers and Chen 1982). Biological responses to broodstock diets, such as increased handling tolerance, increased egg survival, and higher vitamin levels in the eggs at spawning, have been reported when broodstock salmonids are fed carotenoid fortified diets. As spawning approaches, the fish stop feeding, so any nutrients deposited in the eggs must come from maternal reserves.

Specialized feeds.   Salmon smolt diets are often modified to increase or decrease the amount of fat in the fish, a factor that may influence survival after hatchery release. Diet modifications to enhance product quality in commercial salmonid aquaculture include carotenoid pigment supplementation to produce salmon-colored flesh, and substitution of more saturated dietary lipid sources, such as animal fat, to alter the fatty acid composition of the fish and improve taste and storage quality. Another example is dietary vitamin E fortification to extend the frozen storage life of the fish product by increasing tissue vitamin E levels and thereby decreasing lipid oxidation. Development of these feeds is in the experimental stage, but the principles involved are well established in domestic animal production.

Other special formulations for salmonids include low-polluting and maintenance feeds. Low-polluting diets are designed to reduce the organic matter and phosphorus levels in hatchery effluent by reducing the levels of indigestible matter and phosphorus in the diet. Maintenance feeds are formulations designed to maintain fish in a healthy state at very low feeding levels, such as those experienced by coldwater hatcheries in the winter. Another use for a maintenance feed is with coho salmon, which have a target release size in late spring. These fish can easily exceed their release size if fed at normal levels, so their feeding level is reduced during the fall and winter. Maintenence feeds with slightly higher vitamin and trace element levels and lower protein and energy levels than production diets may bring the fish through winter in better shape.

FEEDING PRACTICES

Feed is the largest single operating expense in salmonid aquaculture, averaging about 55% of the cost of rearing fish. Feeding practices, such as feeding rate, feeding frequency, feed particle size, and methods of delivering feed to the fish all affect fish growth rates, food conversion rates, and uniformity of fish size within a pond. These feeding practices should therefore be designed to fit the specific goals of various aquaculture programs.

Feeding rate can be to satiety or at a fixed rate, expressed as a percentage of body weight per day. Feeding rate is affected by temperature, fish size, photoperiod, moisture and energy content of the diet, rearing density, and water quality. Reliable feeding charts have been developed for salmonids which allow near maximum growth rate (Table 9.8). Methods are available for the fish feeder to calculate feeding rate, such as the method of Buterbaugh and Willoughby (1967). By this method, the percentage of body weight to feed daily at a given temperature is calculated by dividing the hatchery constant by the fish length in inches. The hatchery constant is determined by multiplying 300 times the desired food conversion value times the daily increase in fish length. This method is particularly useful for hatcheries using a feed formula that is constant, but has limited value in hatcheries where the formula changes to reflect changes in feed commodity prices.

The inverse relationship between feeding rate and feed efficiency is well known. As shown in Figure 9.4, there are feeding rates at which feed conversion (g feed/g weight gain) will be the lowest (optimum) and at which growth will be maximum (maximum feeding level). In most feeding situations, feeding at levels for near maximum growth rather than for optimum feed conversion, which is at a lower level, is usually economically desirable (see Figure 9.4).

Table 9.8. RECOMMENDED FEEDING RATE (R) AND FREQUENCY (F) FOR SALMON

Temperature (°C)	Fish size (g)															
	0.5–1.5		1.5–2.5		2.5–3.5		3.5–5.0		5.0–7.5		7.5–11.5		11.5–18.0		>18.0	
	R[1]	F[1]	R	F	R	F	R	F	R	F	R	F	R	F	R	F
2.2	2.8	7/5	2.4	7/4	1.9	7/2	1.8	6/1	1.4	5/1	1.4	E/1				
3.3	3.0	7/5	2.6	7/4	2.1	7/2	2.0	6/1	1.7	5/1	1.8	E/1				
4.4	3.4	7/5	2.8	7/4	2.3	7/2	1.9	7/1	1.6	6/1	1.3	5/1				
5.5	3.8	7/5	3.0	7/4	2.5	7/2	2.1	7/1	1.9	6/1	1.4	5/1	1.4	E/1	1.0	E/1
6.5	4.2	7/5	3.3	7/2	2.7	7/2	2.1	7/1	2.1	6/1	1.7	5/1	1.8	E/1	1.2	E/1
7.7	4.6	7/5	3.7	7/4	2.9	7/2	2.5	7/1	2.3	6/1	2.0	5/1	2.0	E/1	1.4	E/1
8.8	5.0	7/5	4.1	7/4	3.2	7/2	2.7	7/1	2.6	6/1	2.2	5/1	2.4	5/1	1.6	E/1
9.9	5.6	7/5	4.5	7/4	3.6	7/2	2.9	7/1	2.8	6/1	2.1	6/1	1.8	5/1	1.8	E/1
11.1	6.2	7/5	4.9	7/4	4.0	7/2	3.2	7/1	3.0	6/1	2.3	6/1	2.1	5/1	2.2	E/1
12.1	6.8	7/5	5.4	7/4	4.4	7/2	3.6	7/1	3.3	6/1	2.6	6/1	2.4	5/1	2.6	E/1
13.2	7.5	7/5	6.0	7/4	4.8	7/2	4.0	7/1	3.7	6/1	2.8	6/1	2.7	5/1	3.0	E/1
14.3	8.3	7/5	6.6	7/4	5.3	7/2	4.4	7/1	4.2	6/1	3.0	6/1	2.9	5/1	3.4	E/1
15.4	9.1	7/5	7.2	7/4	5.9	7/2	4.8	7/1	4.7	6/1	3.3	6/1	3.2	5/1	3.8	E/1

[1] R means daily feed allowance in g/100 g fish weight. F means days fed weekly/number of feedings daily.
Source: Adapted from: Moore-Clark Company, Inc., LaConnor, Washington.

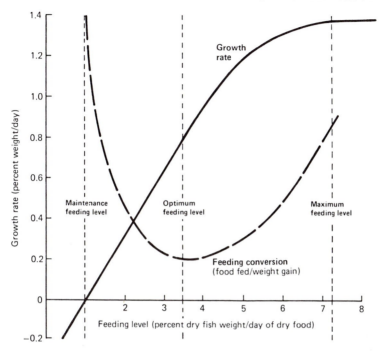

Fig. 9.4. Theoretical relationship between feeding level, growth rate, and food conversion.

Feeding frequency varies with fish size and with feeding level. Rapidly growing salmon can consume between 2.5% and 3.0% of their dry body weight in dry food at a single feeding. Converting this to wet weight basis, salmon can consume 0.75% of their wet body weight of dry feed, 0.82% of 10% moisture feed, or 1.0% of 30% moisture feed at a single feeding. A daily ration should be divided into feedings that provide appropriately sized meals. Thus, if a pond of salmon is being fed moist feed (OMP) at 4% of body weight per day, this ration should be fed in 4 to 5 feedings. If fish are fed too infrequently, the size of the meal may exceed the amount the fish can consume, resulting in wasted feed, decreased water quality, and increased food conversion values. If fish are fed too frequently, increased size variation may occur.

Feed particle size should increase as fish grow, as indicated in Table 9.9. Feeding a feed particle that is too small will result in wasted feed and may increase nutrient leaching from the particle because of increased particle surface area. Feeding a feed particle that is too large will increase feed waste because the fish must allow the particle to break up before swallowing, which also wastes feed.

Table 9.9. RECOMMENDED DRY FEED
PARTICLE SIZES FOR SALMON

Fish weight (g)	Particle size (mm)
< 0.5	0.5
0.5–0.9	0.8
0.9–2.3	1.1
2.3–4.5	1.6
4.5–6.1	2.3
6.1–9.1	2.3
9.1–22.0	4.7
>22.0	

Source: Fowler and Burrows (1971).

Salmon and trout are fed by hand, machine blower, automatic feeders, or demand feeders. Feeding by hand is labor intensive, but it ensures human contact with the fish and may result in more rapid awareness of aquaculture problems as they develop. Automatic feeders are labor-saving devices that are helpful if they are properly managed. They do possess a potential for feed waste, because they deliver the feed regardless of the appetite of the fish. Automatic feeders must be closely monitored both by observation and by frequent food conversion determinations to ensure their effective use. Demand feeders are devices that deliver a small amount of feed in response to an action by the fish. With most demand feeders, a rod extends from the feeder into the water and fish bump the rod causing feed to fall into the water. As long as the delivery system is properly adjusted, little feed wastage occurs. Feeding is spread throughout the day which generally allows for a higher rate of consumption by all fish in the culture system.

REFERENCES

BUTERBAUGH, G. L., and H. WILLOUGHBY. 1967. A feeding guide for brook, brown, and rainbow trout. *Prog. Fish-Cult.* 29: 210–215.

CASTELL, J. O. 1979. Review of Lipid Requirements of Finfish. *Proceedings of the World Symposium on Finfish Nutrition and Fishfeed Techniques.* Hamburg, 20–23 June 1978. Vol. 1. Berlin: Heenemann Verlagsgesellschaft.

FOWLER, L. G., and R. E. BURROWS. 1971. The Abernathy salmon diet. *Prog. Fish Cult.* 33: 67–75.

HILTON, J. W., and S. J. SLINGER. 1981. Nutrition and feeding of rainbow trout. *Can. Spec. Publ. Fish. Aquat. Sci.* 55.

LALL, S. P. 1979. Minerals in finfish nutrition. *Proceedings of the World Symposium on Finfish Nutrition and Fishfeed Techniques.* Hamburg, 20–23 June 1978. Vol. 1. Berlin.

MEYERS, S. P., and H. M. CHEN. 1982. Astaxanthin and its role in fish culture. Proceedings of the Warmwater Fish Culture Workshop. World Maniculture Society Special Publication No. 3: 153–165.

NATIONAL RESEARCH COUNCIL. 1981. *Nutrient Requirements of Coldwater Fishes.* National Academy of Sciences. Washington, DC.

OGINO, C. 1980. Requirements of carp and rainbow trout for essential amino acids. *Bull. Jpn. Soc. Sci. Fish.* 46: 171–174.

WATANABE, T. 1982. Lipid nutrition in fish. *Comp. Biochem.* 73B: 1–15.

WATANABE, T., T. TAKEUCHI, and C. OGINO. 1979. Studies on the sparing effect of lipids on dietary protein in rainbow trout (*Salmo gairdneri*). *Proceedings of the World Symposium on Finfish Nutrition and Fishfeed Techniques.* Hamburg, 20–23 June 1978. Vol. 1. Berlin: Heeneman Verlagsgesellschaft.

10

Practical Feeding—
Penaeid Shrimps

Chhorn Lim and Amber Persyn

Demand for shrimp in the world market, especially in the industrialized countries, has increased considerably in recent years as the oceans have become a less reliable supply. The ocean fisheries for shrimp are at near maximum sustainable yield. Shrimp catch from the sea is unpredictable due to uncontrollable natural phenomena. Pollution and other human activities have disrupted the ecology of shrimp nursery grounds in many areas. Energy requirements for harvesting ocean shrimp is high; diesel fuel represents more than 50% of the total production cost. Therefore, the increase in world consumption of shrimp is likely to be dependent on increases in production through aquaculture. In 1986, pond-raised shrimp represented an estimated 6% to 8% of the world shrimp consumption.

A traditional method of shrimp culture by trapping and holding postlarvae which accidentally enter milkfish ponds through the incoming tidal water has been practiced for centuries in Southeast Asian countries. However, the development of modern shrimp culture technology began in the 1930s when Hudinaga first successfully reared larvae of *Penaeus japonicus* in captivity. Since then, methods have been developed for mass production of marine shrimp larvae using wild gravid females. Within the past 15 years, several shrimp species of the penaeid family, such as *Metapenaeus ensis, Penaeus japonicus* (Figure 10.1), *P. monodon, P. indicus, P. merquiensis, P. aztecus, P. stylirostris, P. setiferus, P. schmitti, P. orientalis,* and *P. vannamei* have been spawned and their larvae reared in captivity. The successful development of this technology is the primary reason that commercial shrimp farming has emerged as an economically successful aquaculture husbandry. Culture methods have evolved from traditional extensive to semi-intensive and intensive systems utilizing sophisticated

Fig. 10.1. The marine shrimp *Penaeus japonicus* is a popular culture species in Japan and other subtropical areas of Asia.

facilities, equipment, and management techniques. However, if high production rates are to be achieved, satisfactory feeds and feeding practices are essential.

CULTURE PRACTICES

The first limiting factor in culture of penaeid shrimp is the availability of good quality seeds for stocking. Wild caught postlarvae constitute the primary source of seed stock in some countries, such as Ecuador, Honduras, Panama, Indonesia, and the Philippines. The quantity of wild seed is unpredictable and fluctuates considerably depending on the ecology and productivity of the nursery grounds, weather conditions, and other factors. Hatchery production now supplies a significant amount of the seed stock, especially in Japan and Taiwan, and will undoubtedly replace wild collection of postlarvae in all areas.

Larval rearing. One of the limiting factors in hatchery production of postlarval shrimp is the availability of spawners. Until the mid-1970s, the spawners used in hatcheries were gravid females taken from the wild. They were in advanced stages of reproductivity and most spawned the day they were captured. At present, several species of penaeid shrimps can be induced to mature and spawn in captivity. The females may be grown in culture ponds or caught in the ocean.

Fertilized eggs hatch into nauplii within 24 hours at a water temperature of 28° C to 30° C. Newly hatched nauplii do not feed and are nourished by the yolk nutrients. The nauplii metamorphose into protozoa after five to six moltings within 48 hours. At this stage the larvae are fed primarily with planktonic diatoms such as *Skeletonema* sp., *Tetraselnius* sp., and *Chaetoceros* sp. Artificial feeds, such as egg yolk powder and yeast, are sometimes given as a supplement to the natural foods. The zoeal larvae have no reflex to search for food but must wait until, by chance, suitable food particles come in contact with the mouth. Thus, a sufficient quantity of food must be kept in suspension in the water in the culture tank at this time.

The zoea molt two to three more times within 4 to 5 days before transforming into myses. Mysis larvae resemble young shrimp, but they swim in a vertical position with head and tail downward. They are fed mainly with *Artemia* nauplii or zooplankton, especially *Brachionus* sp., in addition to the phytoplankton.

The myses metamorphose to postlarvae after three moltings within 3 to 4 days. During the first 5 days of the postlarval stage, they are usually fed with *Artemia*. Artificial feeds, such as small dry diet particles, microencapsulated diets, and minced fish flesh, are offered as substitutes for live foods as the larvae gradually acquire the habit of living and feeding on the bottom of the tanks. The larvae are stocked in nursery or production ponds, depending upon the management practice, after 15 to 20 days.

Nursery practices.   Growing shrimp postlarvae to juveniles in nursery ponds prior to stocking in production ponds offers the advantage of having larger size shrimp to stock. By stocking larger size shrimp, the culture period could be shortened or larger size shrimp can be harvested. Survival rate is higher when postlarvae are grown in nursery ponds than when postlarvae are stocked directly into production ponds. Estimation of feed allowances and harvest yields is easier due to more accurate assessment of survival rate when juveniles (from nursery ponds) are stocked. Disadvantages of using nursery systems are that juveniles are more difficult to handle than postlarvae and higher mortality may result during transport. Additional land and capital are required for the construction and operation of nursery units.

Most modern, large-scale shrimp farms have incorporated nursery systems into the production scheme. The stocking rates used vary considerably depending on the systems and the management practices and the species cultured. Stocking rates in earthen ponds are from 50 to 200 postlarvae/m^2. Higher stocking rates are used for intensive tank systems, ranging from 500 to 1,000 postlarvae/m^2. The postlarvae are fed two to four times daily with dry diets or fresh foods such as chopped fish and mussel. Daily feeding rates vary considerably. When first placed in the nursery ponds, the small (0.1 g) postlarvae are fed 25% to 50% of their weight, divided into four or more daily feedings. As they grow, feed allowance and frequency decrease. Toward the end of the 30- to 50-day nursery period, the feed allowance may decrease to 10% or less of shrimp weight, depending upon the weight of shrimp in the pond. Final average shrimp weight in nursery ponds is about 1 g.

Growing shrimp to harvestable size.   There are great differences in the culture systems and management procedures being used in shrimp farming. These differences are mainly attributed to the availability and cost of land, seedstock, feed, electricity, fuel, and culture technology, and value of the shrimp produced. The methods of shrimp culture may generally be classified under three categories: extensive, semi-intensive, and intensive techniques.

Most shrimp are produced in earthen ponds by extensive or semi-intensive management techniques. Extensive culture is characterized by low stocking density, usually less than 2.5 postlarvae or juveniles/m^2. Natural stocking with postlarvae coming in during high tides is still being practiced in some Asian countries. Supplementary feeding is seldom practiced and the shrimps depend mainly on the natural foods available in ponds. Organic and inorganic fertilizers are used to increase the productivity of natural foods. Water management is done through tidal fluctuation. The yields obtained generally range from 150 to 500 kg/hectare/crop.

In semi-intensive culture operations (see Figure 10.2), the stocking rates used per square meter vary from 3 to 10 juveniles, or up to twice this number if early postlarvae are stocked. Commercial feeds are given as supplements to the natural foods. Fertilizers are also applied to enhance the growth of natural food organisms. Water is pumped through the ponds at a rate 2% to 10% of the pond volume daily when the pond receives feed. Generally, the yields range from 600 to 1,200 kg/hectare/crop, with 2 to 2.5 crops per year.

Fig. 10.2. Semi-intensive shrimp culture pond in Honduras. Size of the pond is 25 hectares; stocking density is 3 to 10 shrimp/m^2; yield is 600 to 1,200 kg/ha/crop; 2 to 2.5 crops/yr are produced; water exchange rate is 2% to 10% of the pond volume per day.

In areas where land cost is very high, intensive culture practices in ponds or tanks may be economically feasible provided there are available technology and high market prices for the product (see Figure 10.3). Intensive culture operations require sophisticated management techniques and nutritionally complete diets. In intensive pond culture, the shrimps are stocked at a rate of $20/m^2$ to $40/m^2$. Water is exchanged daily at rates up to 50% and aeration is usually provided. The shrimps depend primarily on the commercial feed as their source of nutrients; thus, nutritionally complete, concentrated feeds are used. The production ranges from 2,000 to 9,000 kg/hectare/ crop. This culture method is widely practiced in Taiwan and is rapidly gaining acceptance in other parts of the world. In intensive tank cultures, stocking density goes up to 160 juveniles/m^2. High water exchange rate (100–300% daily), continuous aeration, and nutritionally complete, concentrated feeds are used. Production can reach as high as 24,000 kg/hectare/crop. Intensive tank culture techniques are practiced in Taiwan and Japan where shrimp market prices are high.

Fig. 10.3. Intensive shrimp culture pond in El Salvador. Size of the pond is 1 hectare. Aeration is continuous and water exchange rate is about 25% of the pond volume daily. Stocking density is 30 to 40 shrimp/m^2; yield is 6,000 to 8,000 kg/hectare/crop; 3 to 3.5 crops/yr are produced.

NUTRIENT REQUIREMENTS

Proteins and amino acids.   Shrimps, like fishes and other animals, do not have an absolute requirement for protein per se, but require a balanced mixture of indispensable and dispensable amino acids. However, shrimp utilize mixtures of individual amino acids more poorly than amino acid combinations in intact proteins. The optimum dietary protein level for growth of penaeid shrimps has been reported to range from 28% to 60% (Table 10.1). Reasons that these values differ are due to species, size, protein quality, level of nonprotein energy, feeding rate, and availability of natural food organisms. Most of the values in Table 10.1 were determined in absence of natural aquatic foods.

All shrimp species that have been studied require the same 10 essential amino acids as do finfishes and terrestrial animals. Arginine, histidine, isoleucine, leucine, lysine, methionine, phenylalanine, threonine, tryptophan, and valine have been found to be essential for *P*.

Table 10.1. OPTIMUM DIETARY PROTEIN REQUIREMENTS FOR GROWTH OF PENAEID SHRIMPS

Species	Protein source	Protein requirement (%)	Reference
Penaeus japonicus	Shrimp meal Casein	40 45–55	Balazs et al. 1973 Teshima and Kanazawa 1984
	Casein and egg albumin Squid meal	52–57 60	Deshimaru and Yone 1978a Deshimaru and Shigueno 1972
Penaeus monodon	Casein and fish meal Squid meal, shrimp meal, fish meal, soybean meal, and casein	45–50 40–45	Lee 1971 Alava and Lim 1983
Penaeus indicus *Penaeus aztecus*	Prawn meal Soy flour, menhaden meal, and shrimp meal	43 51	Colvin 1976 Zein-Eldin and Corliss 1976
Penaeus merguiensis *Penaeus setiferus* *Palaemon serratus*	*Mytilus edulis* meal Fish meal Shrimp meal, fish meal	34–42 28–32 40	Sedgwick 1979 Andrews et al. 1972 Forster and Beard 1973
Penaeus stylirostris and *Penaeus vannamei*	Fish meal, shrimp meal, and soybean meal	30–35	Colvin and Brand 1977

japonicus (Kanazawa and Teshima 1981), *P. azteus* (Shewbart, Mies, and Ludwig 1972), *Palaemon serratus* (Cowey and Forster 1971), and *P. monodon* (Coloso and Cruz 1980). However, the quantitative requirements for the essential amino acids have not yet been determined, due in part to the poor ability of shrimps to utilize mixtures of crystalline amino acids (Deshimaru and Kuroki 1974).

Generally, the essential amino acid profile of the protein in an animal's body closely approximates the dietary amino acid requirements of the animal. Table 10.2 shows the essential amino acid content of *P. japonicus* muscle, and also of clam, squid, and whole egg protein, which are high quality proteins for shrimps. A shrimp diet with an essential amino acid profile similar to that of *P. japonicus* muscle in the table would likely provide good growth in the fed shrimp.

Casein, which is generally used as a standard protein source for nutrient requirement studies of finfishes and other vertebrates, has been found to be poorly utilized by several shrimp species. This is possibly due to the low arginine content of casein. Fish meal, a high quality protein source for finfishes, seems to have lower nutritional value for shrimp, especially when fed as the sole protein source. This has been reported for *P. monodon* (Destajo 1979), *P. duorarum* (Sick and Andrews 1973), *P. indicus* (Colvin 1976), *Palaemon serratus* (Forster and Beard 1973), and *P. japonicus* (Shigueno 1975). Shigueno (1975) has suggested that this may be due to a shortage of phenylalanine and the basic amino acids (arginine, histidine, and lysine) in fish meals. Soybean meal is a relatively good protein source for shrimp. It provided better growth of *P. duorarum* than fish meal, shrimp meal,

Table 10.2. ESSENTIAL AMINO ACID CONTENT OF PROTEINS FROM SHRIMP (*P. JAPONICUS*) MUSCLE, CLAM, SQUID, AND WHOLE EGG

Amino acid	Proteins			
	Shrimp muscle (%)	Clam (%)	Squid (%)	Whole-egg (%)
Arginine	4.76	4.50	6.99	5.45
Histidine	1.66	1.27	1.45	1.71
Isoleucine	2.89	2.00	2.72	3.46
Leucine	7.04	4.01	5.37	6.47
Lysine	7.24	4.68	4.29	5.45
Methionine	2.92	1.70	2.61	3.01
Phenylalanine	3.90	2.13	3.01	4.15
Threonine	3.62	2.81	3.84	3.73
Tryptophan	0.52	0.51	1.03	0.79
Valine	2.87	2.18	2.49	3.86

Source: Deshimaru 1982.

casein, or corn gluten meal (Sick and Andrews 1973). Combinations of soybean meal and fish meal are commonly used in commercial shrimp feeds.

Supplementing amino acid deficient diets with crystalline amino acids has not been successful with shrimps. Basal casein-albuminin diets fortified with essential amino acids to provide amino acid composition similar to that of diets containing clam protein failed to provide growth of *P. japonicus* equal to that of the clam protein diets (Deshimaru 1982). However, when various protein sources were combined to provide an essential amino acid profile similar to that of the diet containing clam protein, the growth rate of shrimps was essentially the same as that obtained by using the diet with clam protein. The inability of shrimps to utilize free amino acids as substitutes for dietary proteins is probably due to differences in the rate of absorption. Deshimaru (1982) showed that the rate of incorporation of free radioactive arginine into muscle protein was less than 1% as compared to incorporation of 90% of the protein-bound arginine. Covalently bonded methionine has been found to improve soybean protein for *P. Vannamei.*

The protein percentage in commercial feeds fed in intensive culture systems is usually 35% or above. That in semi-intensive culture feeds varies, generally from 20% to 35%. Research at the Enrique Ensenot Marine Laboratory in Panama showed that 25% protein feeds were as productive as higher protein feeds in semi-intensive culture ponds, which were also fertilized, containing 5 shrimp per m^2.

Energy.   Limited information is available on the energy requirements of shrimps. It appears that shrimps utilize part of the dietary protein as a source of energy, with carbohydrates and lipids in the diet; however, excessive levels of protein or high protein/energy ratios in the diet usually result in reduction of the growth rate. Some carbohydrate in the diet has shown nutritional benefit for shrimps. Addition of 1.2 parts carbohydrate per 1 part protein to *P. monodon* diets (Bages and Sloanes 1981) or 25% carbohydrate in 45% protein diets for *P. japonicus* (Teshima and Kanazawa 1984) allowed satisfactory growth.

The optimum energy to protein ratios for various species of shrimp at different sizes have not yet been defined. Sedgwick (1979) showed that excessive energy in the diet decreases feed intake and consequently limits the consumption of protein and other essential nutrients. *P. monodon* juveniles (about 1 g) grew best when fed 38% protein diets having energy values (physiological fuel values for mammalians) of 3.2 to 3.6 kcal/g of diet (Lim and Pascual 1979).

Lipids.   Lipids are required in the diets of shrimps not only for their energy value, but as sources of essential fatty acids, fat-soluble vitamins, sterols, and phospholipids. Shrimps appear to have a dietary requirement for fatty acids of the linoleic and linolenic series. Studies on the biosynthesis of fatty acids by *P. japonicus* using labeled acetate or palmitate indicate that most of the radioactive compounds were incorporated into saturated and monosaturated fatty acids such as palmitic (16:0), palmitoleic (16:1 n-7), stearic (18:00), oleic (18:1 n-9), and eicosamonoenoic (20:1 n-9) acids but very little was converted into linoleic (18:2 n-6), linolenic (18:3 n-3), eicosapentaenoic (20:5 n-3) and docosahexaenoic (22:6 n-3) acids (Jones, Kanazawa, and Ono 1979; Kanazawa et al. 1977). This suggests that fatty acids of the linoleic (18:2 n-6) and linolenic (18:3 n-3, 20:5 n-3 and 22:6 n-3) series are probably dietary essential for *P. japonicus*. Results of other feeding experiments have shown that linoleic (n-6) and linolenic (n-3) fatty acids are also dietary essentials for *P. indicus* and *Palaemon serratus*. Fed alone, linolenic acid was found to be nutritionally superior to linoleic acid, while ecosapentaenoic (20:5 n-3) or docosahexaenoic (22:6 n-3) acids promoted better growth than linolenic acid. Shrimps appear to have limited ability to desaturate and elongate 18:3 n-3 to 20:5 n-3 and 22:6 n-3 fatty acids, which are the biologically active fatty acids. The optimum dietary levels of 20:5 n-3 or 22:6 n-3 fatty acids for shrimps are estimated to range from 0.5% to 1.0%. The optimum ratio of 18:3 n-3/18:2 n-6 seems to be about 1.2 for *P. serratus* (Martin 1980) and *P. stylirostris* (Fenucci, Lawrence, and Zein-Eldin 1981).

Crustaceans do not synthesize sterols from acetate or mevalonate as do finfish, and therefore they require a dietary source. Cholesterol is the major sterol found in crustaceans and is a precursor of sex hormones, molting hormones, and a constituent of the hypodermis in crustaceans. Minimum dietary level of cholesterol required by shrimps have been reported to be 0.5% to 1.25%. *P. japonicus* can partially substitute phytosterols and fungal-sterols, such as ergosterol, stig-masterol, and beta-sistosterol, for dietary cholesterol. These sterols are converted metabolically to cholesterol.

In addition to the essential fatty acids and sterols, crustaceans seem to also have a dietary requirement for phospholipids. Supplementation of 1% lecithin into diets of *P. japonicus* juveniles and larvae, especially, improved growth and survival. Lecithin from clam (*Tapes* sp.), soybean, bonito egg, and phosphatidylinositol from soybean were suitable sources of phospholipid. Effective phospholipids for shrimps appear to be those containing choline or inositol and polyunsaturated fatty acids. Finfishes do not have a dietary requirement for preformed phospholipids. The reason that some shrimps have this requirement is apparently due to the slow rate of biosynthesis or the high metabolic requirement for phospholipids for transport of dietary lipids, especially cholesterol, in the hemolymph.

Carbohydrates. Shrimps utilize dietary carbohydrates as energy sources. Polysaccharides, such as starch and dextrin, are better energy sources than the monosaccharide, glucose. Glucose in the diet at levels of 10% or above suppresses growth.

The exoskeleton of crustaceans, which is shed repeatedly during growth, contains the glucosanine, chitin. Evidence on the essentiality of glucosamine in shrimp diets is conflicting. Some studies showed that dietary supplementation of glucosanine improved the growth of *P. japonicus* while others failed to show an improvement.

Vitamins. *P. japonicus* requires vitamin E, vitamin D, beta-carotene (vitamin A activity), thiamin, riboflavin, pyridoxine, vitamin B_{12}, folic acid, nicotinic acid, biotin, choline, inositol, and vitamin C in the diet for normal growth (Kanazawa 1985). Information on the quantitative requirements is limited. The dietary levels of various vitamins that have been reported for shrimps are considerably higher than those found for finfishes. For example, reported requirements are 1,000 to 10,000 mg/kg for vitamin C, 60 to 120 mg/kg for thiamin, 120 mg/kg for pyridoxine, and 2,000 to 4,000 mg/kg for inositol. Whether or not shrimps actually have a metabolic requirement for such high levels of vitamins or whether a substantial quantity of vitamins is lost into the water during ingestion by the shrimp is unknown. In the absence of clear-cut information on the vitamin requirements of shrimps, vitamin allowances for finfish may be used as guides. However, the amounts of vitamins used in shrimp feeds, especially purified diets of feeds for intensive culture should be higher than those used for finfish to allow for losses during ingestion.

Shrimp fed diets deficient in vitamin C developed "black death" syndrome, which is characterized by blackened lesions in the subcuticular tissues of the body surface; in the walls of the esophagus, stomach, and hindgut; and in the gills and gill cavity (Lightner et al. 1977). Vitamin C deficiency signs in *P. japonicus* are discoloration and development of a greyish-white color on the margin of the carapace, the lower part of the abdomen, and on the tips of the walking legs (Deshimaru and Kuroki 1976).

Minerals. Shrimps, like other aquatic species, absorb several minerals from the surrounding seawater. *P. japonicus* juveniles did not have a dietary requirement for calcium, magnesium, and iron, but required phosphorus, potassium, and trace minerals in the diet (Deshimaru and Yone 1978b). Although penaeid shrimps do not have a dietary requirement for calcium, this mineral is usually included in the diet to maintain a ratio of approximately 1 to 2 parts calcium to 1 part phosphorus. Kanazawa (1982) reported that the optimum levels of minerals in the diet of *P. japonicus* juveniles are 1% calcium, 1% phosphorus, 0.9% potassium, 0.3% magnesium, and 0.006% copper.

FEEDS AND FEEDING

Natural foods. Penaeid shrimps are regarded as omnivorous scavengers that feed on a variety of benthic organisms and detritus, but they cannot be placed in any one trophic level because they are generally opportunistic feeders. The food habits of shrimps vary during the life stages. At zoea and mysis, the larvae feed on free swimming plankton. The postlarvae, being strictly demersal, are detritivores. The feeding habit of juveniles is at first omnivorous and then changes to carnivorous, and they prey mainly on slowly moving microinvertebrates. Adult penaeid shrimps are opportunistic feeders, but seem to prefer foods of animal rather than plant origin. Small crustaceans, molluscs, fish, polychaetes, and annelids constitute the principal natural diet components of shrimps. Under pond conditions, the primary source of natural foods for shrimps is the thin aerobic layer of the pond bottom. This layer consists of both living and dead algae and plankton, bacteria, detritus, and other benthos such as polychaetes and annelids. Bacteria have been found to comprise 10% to 20% of the total organic carbon in the stomach contents of several species of shrimps (Moriarty 1977).

Feeding behavior. Shrimps find their foods mainly by chemosensory mechanisms rather than vision. The chemoreceptors are concentrated on the anterior appendages, antennae, and antennules. Once the scent is detected, the shrimps become alerted, move over the substrate toward the direction of the food, and rapidly seize the food with either of the first three pairs of the chelate pereiopods. Each pereiopod can work separately in either locating, gathering, holding, or conveying the food to the mouth parts. The mouth parts, which are comprised of three pairs of maxillipeds, two pairs of maxillae, and a pair of heavily chitinized mandibles, act together to reduce large food particles to a size suitable for ingestion. This apparently is an opportunity for loss of nutrients from processed feeds to the water.

The ingested food is further chewed to fine particulate size by the mandibles before being swallowed. The food passes through the esophagus and enters the anterior chamber of the proventriculus (foregut), where it is further reduced to a semifluid state mechanically and by digestive enzymes, and separated into fluid and coarse fractions by dense setae. The fluid passes into the posterior chamber and finally into the tubules of the hepatopancreas for further digestion and absorption. The coarser particulates pass directly to the midgut, where there is some digestion, and which is an important site of nutrient absorption. The undigested and unabsorbed portions of food enter into the hindgut, which serves mainly as a region for the compaction and transportation of fecal materials.

PRACTICAL FEEDS

Traditionally, cultured shrimps were fed fresh or frozen trash fish, mussel, clam, or squid. Presently, these food items are used mainly for broodstock and occasionally at postlarval stages. Live foods, such as algae, rotifers, and *Artemia,* are still major sources of food for shrimp larvae although various types of artificial diets substitute for some of the live food.

Commercially processed feeds are successfully used in nurseries and semi-intensive and intensive grow-out operations. Many commercial shrimp feeds are available worldwide. Although research information is available on basic nutrient requirements of several shrimp species, there is a scarcity of research data on recommendations for pond feeds. Because the culture environment makes a valuable contribution to the nutrient requirements of shrimp, cost-effective feeds for the various culture systems and management practices are difficult to design. The necessity of supplementation of various pond feeds with all nutrients, such as vitamins and essential lipids, has not been established. Most commercial feeds contain a vitamin premix. Generally, higher protein diets are fed during early postlarval stages and juvenile stages, and the protein percentage decreases during the grow-out period. Examples of formulas for practical shrimp diets are given in Table 10.3.

Due to the benthic feeding behaviors of shrimps, practical commercial feeds should be processed into sinking pellets. Sizes of the pellets vary depending on size of the shrimps. Crumbles are used during postlarval stages and pellets are fed from juvenile through marketable

Table 10.3. MODEL FORMULAS OF PRACTICAL SHRIMP FEEDS, FOR INTENSIVE
CULTURE (38% PROTEIN) AND SEMI-INTENSIVE CULTURE (30 AND
25% PROTEIN)

	Protein percentage		
Ingredient	38%	30%	25%
Fish meal	16	12	10
Shrimp head meal	15	10	10
Squid meal	5	—	—
Soybean meal	30.8	28.6	20.8
Cereal products or by-products[1]	22–24	39–41	49–51
Fish oil	4	4	4
Soybean lecithin	1	—	—
Cholesterol	0.2	—	—
Binder	1–3	1–3	1–3
Dicalcium phosphate	2.3	2.7	2.9
Vitamin mix[2]	0.5	0.5	0.5
Trace mineral mix[3]	0.05	0.05	0.05

[1] Cereal products or by-products rich in starch, such as ground whole wheat, high-gluten
wheat flour, and corn meal are recommended.
[2] A vitamin premix for warmwater fish feeds should be used, such as that used in
channel catfish feeds in Table 7.3. The allowance should be 50% higher for the
intensive culture feed.
[3] A trace mineral premix, similar to that used in Table 7.3, should be used.

size. Recommended pellet diameter for various sizes of shrimps is
given in Table 10.4.

Good water stability is very important for shrimp feeds (see Figure
10.4). While finfishes usually swallow the whole feed pellet once they
have learned to accept the feed, shrimps are selective and slow eaters.
They take the feed with the chelate pereopods, then convey it to the
mouth parts, which act together to reduce the size of feed to small
particles prior to ingestion. Thus, shrimp pellets should remain stable

Table 10.4. RECOMMENDED PELLET DIAME-
TER FOR VARIOUS SIZES OF SHRIMPS

Stage/size	Particle diameter (mm)
P_{15}–P_{30}[1]	<0.5
P_{30}–0.5 g	0.5–0.8
0.5 g–2.0 g	1–2
2.0 g–5.0 g	2
5.0 g–10.0 g	2–3[2]
>10.0 g	3–4

[1] P means postlarvae and the number represents
the number of days after the larvae have
reached the postlarval stage.
[2] Pellet with 2 mm diameter and 2–3 mm in
length.

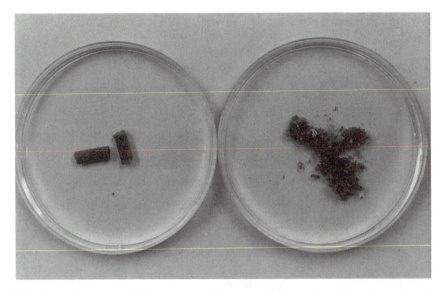

Fig. 10.4. Water stability is extremely important for shrimp feeds because of the slow feeding behavior of shrimp. The pellets above were held in water for 5 minutes; those on the left will remain intact for several hours in water until consumed by the shrimp. Those on the right were not processed with good technology and most of their nutrients will not be consumed directly by the shrimp.

in the water for several hours until consumed by the shrimp. Procedures for processing shrimp feeds with good physical properties are discussed in chapter 5.

Because the feeding response of shrimps is mainly chemosensory, attractants in the diets may increase their feeding activity. Various substances, such as amino acids; fatty acids; and extracts of fish, shrimp, squid, mussel, and clam, have been shown to stimulate feeding response in shrimps. These studies have been primarily under laboratory conditions and the value of attractants under pond feeding conditions is not clearly established.

Feeding. Good quality feeds can give poor results unless proper feeding practices, such as amount, frequency, and method of feeding are used. The daily amount of feed offered is affected by the size of shrimps, feeding schedule, stocking density, availability of the natural foods, and water quality. Daily feed allowances for shrimps range from around 25% of body weight for larvae to less than 3% of body weight per day. Some farmers, however, feed their shrimps according to the

Table 10.5 RECOMMENDED FEEDING RATES AND FRE-
QUENCIES FOR VARIOUS SIZES OF SHRIMPS

Stage/size	Daily feeding rate (% of body weight)	Times fed daily
$P_{15}-P_{30}$	30–20	6
$P_{30}-0.5$ g	20–15	4
0.5–2 g	15–12	3–4
2–5 g	12–8	3
5–10 g	8–6	3
10–20 g	6–4	2–3
>20 g	4–3	2–3

demand. At each feeding, some feed is placed in a feeding tray or on a platform and these are checked a few hours after feeding. The feed allowance will be increased or decreased on the basis of the amount of feed remaining in the feeding tray. Some farmers determine the daily feed allowance from the activity and behavior of the shrimp. One or two hours after feeding, if a significant number of shrimp are swimming actively along the pond banks, this is an indication of under-feeding.

Because shrimps are very slow eaters and feed more or less continuously, multiple daily feeding is desirable. Under laboratory conditions, the optimum feeding frequency of *P. monodon* juveniles (2 g average weight) was three times per day (Lim and Pascual 1979). Disintegration of the feed and loss of the water soluble nutrients can be minimized through multiple daily feeding. Recommended feeding rates and frequencies for various sizes of shrimps are given in Table 10.5.

Unlike fish, shrimps are territorial and do not swim great distances to get food. Thus, it is important to distribute the feed uniformly over the pond. For small ponds or where the labor is relatively inexpensive, feeding is done by hand broadcasting. In large ponds, feeds are offered by the use of boats or airplanes.

REFERENCES

ALAVA, V. R., and C. LIM. 1983. The quantitative dietary protein requirements of *Penaeus monodon* juveniles in a controlled environment. *Aquaculture* 30: 53–61.

ANDREWS, J. W., L. V. SICK, and G. J. BAPTIST. 1972. The influence of dietary protein and energy levels on growth and survival of penaeid shrimp. *Aquaculture* 1: 341–347.

BAGES, M., and L. SLOANES. 1981. Effects of dietary protein and starch levels on growth and survival of *Penaeus monodon* (Fabricius) postlarvae. *Aquaculture* 25: 117–128.

BALAZS, G. H., E. ROSS, and C. C. BROOKS. 1973. Preliminary studies on the preparation and feeding of crustacean diets. *Aquaculture* 2(4): 369–377.

COLOSO, R. M., and L. J. CRUZ. 1980. Preliminary studies in some aspects of amino acid biosynthesis in juveniles of *Penaeus monodon Fabricius*. *Bul. Phil. Bioch. Soc.* 3(1 & 2): 12–22.

COLVIN, P. M. 1976. Nutritional studies on penaeid prawns: Protein requirements in compounded diets for juvenile *Penaeus indicus* (Milne Edwards). *Aquaculture* 7: 315–326.

COLVIN, L. B., and C. W. BRAND. 1977. The protein requirements of penaeid shrimp at various life-cycle stages in controlled environment systems. *Proc. Ann. Workshop World Maricul.* Soc. 8: 821–840.

COWEY, C. B., and J. R. M. FORSTER. 1971. The essential amino-acid requirements of the prawn, *Palaemon serratus*. The growth of prawns on diets containing proteins of different amino-acid compositions. *Mar. Biol.* 10: 77–81.

DESHIMARU, O. 1982. Protein and amino acid nutrition of the prawn, *Penaeus japonicus. Proceedings of the Second International Conference in Aquaculture Nutrition: Biochemical and Physiological Approaches to Shellfish Nutrition,* 106–122 October 17–19, 1981, Rehoboth Beach, DE.

DESHIMARU, O., and K. SHIGUENO. 1972. Introduction to the artificial diet for prawn *Penaeus japonicus. Aquaculture* 1(1): 115–133.

DESHIMARU, O., and K. KUROKI. 1974. Studies on a purified diet for prawn. I. Basal composition of diet. *Bul. Jap. Soc. Sci. Fish.* 40(4): 413–419.

DESHIMARU, O., and K. KUROKI. 1976. Studies on a purified diet for prawn. VIII. Adequate dietary levels of ascorbic acid and inositol. *Bul. Jap. Soc. Sci. Fish.* 42(5): 571–576.

DESHIMARU, O., and Y. YONE. 1978a. Optimum level of dietary protein for prawn. *Bul. Jap. Soc. Sci. Fish.* 44(12): 1395–1397.

DESHIMARU, O. and Y. YONE. 1978b. Requirement of prawn for dietary minerals. *Bul. Jap. Soc. Sci. Fish.* 44(8): 907–910.

DESTAJO, W. H. 1979. Fish meal and shrimp meal as major protein sources for *P. monodon* Fab. juveniles. Master's thesis, University of the Philippines System, Philippines.

FENUCCI, J. L., A. L. LAWRENCE, and Z. P. ZEIN-ELDIN. 1981. The effects of fatty acid and shrimp meal composition of prepared diets on growth of juvenile shrimp, *Penaeus stylirostris. J. World Maricult. Soc.* 12(1): 315–324.

FORSTER, J. R. M., and T. W. BEARD. 1973. Growth experiments with the prawn *Palaemon serratus* Pennant fed with fresh and compounded foods. *Fish. Invest.,* London, Ser. 2, 27(7): 1–16.

JONES, D. A., A. KANAZAWA, and K. ONO. 1979. Studies on the nutritional requirements of the larval stages of *Penaeus japonicus* using microencapsulated diets. *Marine Biology* 54: 261–267.

KANAZAWA, A. 1982. Penaeid nutrition. *Proceedings of the Second International Conference in Aquaculture Nutrition: Biochemical and Physiological Approaches to Shellfish Nutrition,* 87–105 October 27–29, 1981. Rehoboth Beach, DE. 87–105.

KANAZAWA, A.   1985.   Nutrition of penaeid prawns and shrimps. *Proceedings of the First International Conference on the Culture of Penaeid Prawns/Shrimps.* Iloilo City, Philippines, 1984. SEAFDEC Aquaculture Department: 123–130.

KANAZAWA, A., and S. TESHIMA.   1981.   Essential amino acids of the prawn. *Bul. Jap. Soc. Sci. Fish.* 47(10): 1375–1379.

KANAZAWA, A., S. TOKIWA, M. KAYAMA, and M. HIRATA.   1977.   Essential fatty acids in the diet of prawn. I. Effects of linoleic and linolenic acids on growth. *Bul. Jap. Soc. Sci. Fish.* 43(9): 1111–1114.

LEE, D. L.   1971.   Studies on the protein utilization related to growth in *Penaeus monodon* Fabricius. *Aquaculture* 1(4): 1–13.

LIGHTNER, D. V., L. B. COLVIN, C. BRAN, and D. A. DONALD.   1977.   Black death, a disease syndrome of penaeid shrimp related to a dietary deficiency of ascorbic acid. *Proc. Ann. Workshop World Maricul.* 8: 611–624.

LIM, C., and F. P. PASCUAL.   1979.   Nutrition and feeding of *P. monodon* in the Philippines. *Proc. Techn. Consultation on Available Techn. in the Philippines,* 162–164 February 8–11, 1979, SEAFDEC Aquaculture Department, Tigbauan, Iloilo, Philippines.

MARTIN, B. J.   1980.   Croissance et acides gras de la crevette *Palaemon serratus* (Crustacean, decapoda) nourrie avec des aliments composes contenant differentes proportions d'acide linoleique et linolenique. *Aquaculture* 19: 325–337.

MORIARTY, D. J. W.   1977.   Quantification of carbon, nitrogen and bacterial biomass in the food of some penaeid prawns. *Aust. J. Mar. Freshwater Res.* 28: 113–118.

SEDGWICK, R. W.   1979.   Influence of dietary protein and energy on growth, food consumption and food conversion efficiency in *Penaeus merquiensis* de Man. *Aquaculture* 16: 7–30.

SHEWBART, K. L., W. L. MIES, and P. D. LUDWIG.   1972.   Identification and quantitative analysis of the amino acids present in protein of the brown shrimp, *Penaeus aztecus. Mar. Biol.* 16: 64–67.

SHIGUENO, K.   1975.   Shrimp culture in Japan. Association for International Technical Promotion, Tokyo, Japan.

SICK, L. V., and J. W. ANDREWS.   1973.   The effect of selected dietary lipids, carbohydrates and proteins on the growth, survival and body composition of *P. duorarum. Proc. Ann. Workshop World Maricul.* 4: 263–276.

TESHIMA, S., and A. KANAZAWA.   1984.   Effects of protein, lipid and carbohydrate levels in purified diets on growth and survival rates of the prawn larvae. *Bul. Jap. Soc. Sci. Fish.* 50(10): 1709–1715.

ZEIN-ELDIN, Z. P., and J. CORLISS.   1976.   The effect of protein levels and sources on growth of *Penaeus aztecus.* FAO Technical Conference on Aquaculture, Kyoto, Japan, 26 May–2 June 1976. FIR: AQ/Conf/76/E.33: 1–8.

11

Practical Feeding–Eels

Shigeru Arai

Eels are generally classified as warmwater fish. Three species of *Anguilla* genus are important commercially: the Japanese eel *A. japonica,* European eel *A. anguilla,* and American eel *A. rostrata.* Culture of eels for food is most highly developed in eastern Asia, especially in Japan. Eels have been cultured in Japan since the late 1800s, and production in 1983 was about 35,000 tons. Taiwan started eel culture in about 1968 and the industry has developed rapidly, with production of approximately 30,000 tons in 1985, most of which was exported to Japan. Demand for young eels (elvers) usually exceeds the supply. The supply of elvers depends on catches along the coasts of Japan, Taiwan, Korea, and China, and due to the large fluctuation in elver catches, the price varies greatly.

Because of the shortage of Japanese eel elvers and the low price of imported European eel elvers, Japanese eel farmers have tried to culture the European eel with traditional Japanese eel culture methods. However, this has not been highly successful. The European eel is less tolerant of the high water temperatures in Japan and is highly susceptible to parasites and diseases under Japanese culture conditions. Some farmers have succeeded in culturing European eels in systems with a high rate of water exchange.

Commercial eel culture in Europe and North America is not well established, with most of the production coming from catches in natural waters. In 1985, nearly 90,000 tons of eels were consumed in the world, with a value of approximately one billion dollars. Demand for eel worldwide is high, while catches of eels in natural waters are decreasing yearly; thus, eel culture should continue to increase for many years.

CULTURE METHODS

Most eel culture is intensive, which requires good water quality management techniques. Two types of water systems are used in eel culture, the static pond system and the flow-through pond system. Water quality is easier to maintain in the flow-through system, but the static pond system uses less water and energy for pumping. Underground water is usually used for eel culture, but river water can also be used if pollution is not a problem. Generally, the oxygen content of underground water is low, so aeration is required. The oxygen level must be kept above 1 mg/L.

Water temperature is also an important factor, and successful eel culture requires a suitable temperature for a minimum of 6 months per year. The water temperature range for Japanese eels is 13° C to 30° C, and the optimum temperature for growth is 23° C to 30° C. European eels have a lower temperature range, 8° C to 23° C, and a lower optimum temperature, 20° C to 23° C.

Eel farmers in Japan use several sizes of ponds for various growth stages of the eels. The eels are sorted periodically as they grow and the density in the ponds is decreased. The elver nursery pond is about 30 m^2 to 300 m^2, the second stage (intermediate) rearing pond is about 100 m^2 to 1,000 m^2, and the pond for growing to commercial size is about 300 m^2 to 3,000 m^2. Water depth is usually 40 cm to 50 cm in the nursery pond and 70 cm to 80 cm in the intermediate and commercial ponds. Nursery ponds and some intermediate ponds are usually built in a greenhouse. Concrete ponds are the most common rearing units, but net-pen culture in large ponds or lakes is also used for growing eels to commercial size.

Stocking rate in ponds depends on the supply of water and the oxygen content. About 0.2 kg to 0.3 kg of Japanese eels per m^2 are stocked in static ponds, while the stocking rate is about 10 times higher in flow-through ponds.

Commercial eel farmers in Japan specialize in one of two types of production: short period culture, from glass eel and elver to medium size (50 g); and long period culture, from glass eel and elver to commercial size (150–250 g).

NUTRITIONAL REQUIREMENTS

Vitamins.   The qualitative requirements of young Japanese eels for water soluble vitamins have been determined using a purified vitamin test diet which was formulated for Japanese eel by modifying the vitamin test diet developed for chinook salmon (Arai, Nose, and Hashimoto 1971, 1972). Young Japanese eels require 11 water-soluble vitamins, but not para-aminobenzoic acid. Deficiency signs for these vitamins have been determined and are summarized in Table 2.5 in chapter 2 (page 30). Most of the symptoms disappeared when the missing vitamin was fed. Mortality in vitamin deficient fish was low except among eels fed a pantothenic acid-deficient diet. The symptoms observed in vitamin C-deficient eels were similar to scurvy found in warm-blooded animals, but spinal deformities, like those observed in other vitamin C-deprived fish, were not present in eel. Eels fed either choline-deficient or inositol-deficient diets developed similar discoloration of the intestine. This was observed very quickly, within 2 to 4 weeks with choline deficiency. Both of these vitamins are components of phospholipids and play roles in fat metabolism. Dermatitis, including hemorrhage and congestion of the fins, was induced by deficiencies of riboflavin, pantothenic acid, or niacin. The skin lesions induced by lack of these water soluble vitamins may be responsible for the prevalence of fungal disease in the eels.

Vitamin E is the only fat soluble vitamin that has been evaluated with eels to date. Yamakawa et al. (1975) observed poor appetite, poor growth, hemorrhage and congestion in fins, and dermatitis in alphatocopherol deficient eels. They estimated the minimum requirement for growth in young eels is about 200 mg/kg of dry diet. The other fat soluble vitamins, A, D, and K, are required in diets of other fishes and should be included in eel diets to prevent possible deficiencies.

Minerals.   It is well known that fish can absorb minerals from the water. However, Arai, Nose, and Hashimoto (1971) reported that young Japanese eels require dietary sources of some minerals for growth. Eels fed a purified casein diet without a supplemental mineral mix stopped growing in 2 weeks, followed by a gradual loss of body weight and high mortality. External signs of scoliosis and lordosis were found in eels fed the diet without the supplemental mineral mixture.

Eels fed a high fish meal diet showed improved growth when a mineral supplement was added (Arai, Nose, and Kawatsu 1974). Eels fed the unsupplemented diet had hypochromic microcytic anemia. These observations indicate that the minerals in the fish meal may be insufficient or imbalanced for normal growth and health of eels.

Requirements for calcium, magnesium, phosphorus, and iron have been determined for Japanese eels (Nose and Arai 1976). Eels fed diets deficient in either calcium or phosphorus lost their appetite within

1 week and those fed magnesium-deficient or iron-deficient diets lost their appetite after 3 to 4 weeks. Iron-deficient eels had hypochromic microcytic anemia. The minimum dietary requirements for calcium, magnesium, phosphorus, and iron for young eels are 2,700 mg/kg, 400 mg/kg, 2,500 to 3,200 mg/kg and 17 mg/kg of dry diet, respectively.

Protein and amino acids.   The minimum level of amino acid-balanced dietary protein required by young eels for positive growth is 13% and the minimum level for optimum growth is approximately 45% (Nose and Arai, 1972). European eels and Japanese eels require a dietary source of 10 amino acids: arginine, histidine, isoleucine, leucine, lysine, methionine, phenylalanine, threonine, tryptophan, and valine. The quantitative requirements of young Japanese eels for these amino acids are shown in Table 2.4 in chapter 2 (page 26). Arai, Nose, and Hashimoto (1971) reported that L-cystine, a nonessential amino acid, seems to be a growth promoting factor for eels. They doubled the growth rate of eels by adding 1% L-cystine to a casein-gelatin diet supplemented with 0.5% tryptophan.

Lipids.   Lipids are an important energy source for eels. A 2:1 mixture of corn oil and cod liver oil gave better growth to eels fed a casein diet than did either oil alone (Arai, Nose, and Hashimoto 1971).

The addition of either 18:2 n-6 or 18:3 n-3 fatty acids to eel diets improved growth; 18:3 n-3 was more effective (Takeuchi et al. 1980). However, best weight gain was obtained when the diet contained 0.5% 18:2 n-6 and 0.5% 18:3 n-3. The supplemental effect of 1% n-3 highly unsaturated fatty acids was almost the same as that of 1% n-3 18:3 fatty acid. Thus, the fatty acid requirement of eels can be satisfied with 0.5% of each of 18:2 n-6 and 18:3 n-3, or with 1% of 18:3 n-3 in the diet.

FEED PREPARATION

Moist feeds are usually fed in commercial eel culture. The composition of a typical eel feed used in Japan is shown in Table 11.1. Water and oil are added to the dry ingredient mixture shortly before feeding. Approximately 5 to 10 kg of oil and 80 to 100 kg of water are mixed with 100 kg of the dry mixture. The feed should be used quickly after preparation because the alpha (pregelatinized)-potato starch used as a binder weakens quickly and the diet may disintegrate in the water soon after it is fed.

Table 11.1 COMPOSITION OF MODEL GROWER DIET FOR EEL

Ingredients	Percent	Kg per 100 kg of dry mixture
Dry mixture		
White fish meal (65% protein)	65	
Pregelatinized potato starch	22.5	
Soybean flour (48% protein)	5	
Yeast	3	
Liver meal	2	
Vitamin mixture[1]	0.5	
Mineral mixture[2]	2	
Fish oil		5–10
Water		80–100

[1] Use a complete vitamin supplement for fish, such as that used in Table 7.3 for channel catfish or Table 7.21 or 7.22 for salmon.
[2] Use a trace mineral mixture for fish, such as that presented in Table 7.3 for channel catfish or Table 7.21 or 7.22 for salmon.

The composition of Japanese eel feeds can be modified from that in Table 11.1 using local ingredients and other animals protein sources. The alpha starch can be substituted with other binders, such as alginic acid or carboxymethylcellulose (CMC). Approximately 2% to 3% of alginic acid or 4% to 6% of CMC will allow substitution of the alpha starch, which is an expensive ingredient, with a cheaper carbohydrate source, such as wheat flour.

Although eels are usually fed moist feeds, pelleted dry diets can also be used if they are trained to eat dry pellets from the elver stage. It is important that the pellet size be suitable for the various growth stages of the eels. Frequent sorting of the eels by size is necessary when pellets are used instead of moist food.

FEEDING PRACTICES

Because the life cycle of the eel is not yet controlled, elvers caught in nature should be trained to eat artificial feed or trash fish as early as possible. Traditionally, live tubificid (family *Tubifidae*) worms, beef liver, fresh fish, and short-necked clam have been used as starter feeds for elvers. Tubificid has given the best results. After stocking, the elvers should be fed in a feeding basket that hangs close to the water surface at a fixed feeding place in the nursery pond. Chop the tubificid into small pieces for the first 2 to 3 days. About 1 week is required to train the elvers to eat tubificid at a fixed feeding place. If beef liver is used as a starter diet instead of tubificid, the hard connective tissue is removed and the liver is finely ground. Commercially prepared and frozen feeds for elvers are used successfully in Japan as a substitute for live tubificid. After the elvers are accustomed to eating the starter feed from the basket, they are changed to a production feed by feeding a mixture of the two feeds and gradually decreasing the amount of starter in the mixture over several days.

After conversion to the production feed, the elvers should be fed the amount that they can consume in about 10 to 20 minutes. At first, the elvers should be fed two to four times daily, but when the eels become larger, they will eat sufficiently at one daily feeding.

In commercial eel culture, it is customary to feed late in the morning. It is important, particularly in a static pond system, to feed well after sunrise so that photosynthesis has replaced the dissolved oxygen in the pond that was depleted during the night. Eels are fed the moist feeds in baskets at the water surface (Figure 11.1). Any feed remaining in the basket or on the bottom of the pond should be taken out immediately after the allowed feeding time, even if some eels continue to eat.

DAILY FEEDING RATE

Daily allowance of a commercial feed, on a dry weight basis, at a water temperature of 25° C is 6% to 8% of body weight for elvers and small eels and 2% to 3% for larger eels. For feeding raw fish flesh, the feeding rate on a wet weight basis at 25° C is 20% to 30% of the total wet body weight for elvers and small eels and about 10% for larger eels. Adjustments in daily feed allowance should be made on the basis of observed feeding activity of the eels and water quality conditions.

Usually, growth varies greatly among eels in the same culture system. Eel farmers should sort eels by size periodically during rearing and feed the different size groups separately. Eels may be selectively harvested by topping off the larger ones when they reach marketable size and feeding the remaining ones longer. Adult male eels are harvested at about 40 cm to 50 cm in length or 120 g to 500 g in weight, while females reach 100 cm in length and 1 kg to 2 kg in weight during the same time period. Marketable size for eels usually exceeds 0.5 kg in Europe. This means that Europeans consume mainly females and the Japanese consume both males and females. If a farmer wants to culture large eels for the European market, he or she should select mainly females, but more elvers will be needed. Female eels can reach about 0.5 kg to 1 kg size in 2 to 4 years, but males seldom reach over 500 g.

Fig. 11.1.   Eels are fed moist feeds, in large chunks, in baskets at the surface of the water.

RESEARCH NEEDS

At present, eel culture in Japan depends largely upon feeds that contain 60% to 70% high quality fish meal. With the increasing demand for formulated feeds and the shortage and high price of high quality fish meal, substitutes for fishmeal in eel feeds are necessary. The nutritional requirements of eels are not fully known. More information on this, especially the energy requirements, is needed for making least-cost feeds. The life cycle of the eel is not controlled yet. Due to the fluctuation of the elver supply, eel culture has not yet become a sound industry. Studies on artificial spawning have been carried out in Japan and in other countries and hormone injection techniques have been used to accelerate maturation and spawning. However, artificial larval production is not yet a commercial reality.

REFERENCES

ARAI, S., T. NOSE, and Y. HASHIMOTO.   1971.   A purified test diet for the eel, *Anguilla japonica. Bull. Freshwater Fish. Res. Lab.* 21: 161–178.

ARAI, S., T. NOSE, and Y. HASHIMOTO.   1972.   Qualitative requirements of young eels *Anguilla japonica* for water-soluble vitamins and their deficiency symptoms. *Bull. Freshwater Fish. Res. Lab.* 22: 69–83.

ARAI, S., T. NOSE, and H. KAWATSU.   1974.   Effect of minerals supplemented to the fish meal diet on growth of eel, *Anguilla japonica. Bull. Freshwater Fish. Res. Lab.* 24: 95–100.

NOSE, T., and S. ARAI.   1972.   Optimum level of protein in purified diet for eel, *Anguilla japonica. Bull. Freshwater Fish. Res. Lab.* 22: 145–155.

NOSE, T., and S. ARAI.   1976.   Recent advances in studies on mineral nutrition of fish in Japan. FAO, FIR:AQ/Conf/76/E.25, p. 12.

TAKEUCHI, T., S. ARAI, T. WATANABE, and Y. SHIMMA.   1980.   Requirement of eel *Anguilla japonica* for essential fatty acids. *Bull. Japn. Soc. Sci. Fish.* 46: 345–353.

YAMAKAWA, T. S., S. ARAI, Y. SHIMMA, and T. WATANABE.   1975.   Vitamin E requirement for Japanese eel. *Vitamins* 49: 62. (Abstract in Japanese.)

12

Practical Feeding—Crawfish

Edwin H. Robinson

Freshwater crayfish (family *Astacidae*), or crawfish as they are called commercially, are found throughout the world. There are more than 300 species of crawfish worldwide and more than 100 species in the United States. Crawfish are consumed in the south-central United States, primarily Louisiana, and in many areas of Europe. Two groups are cultured in the United States, *Procambarus* sp. and *Orconectes* sp. The red or red swamp crawfish (*P. clarkii*) (Figure 12.1) and the white or white river crawfish (*P. acutus*) are the primary species cultivated in the United States. The red crawfish is normally predominant, but the white crawfish may occur in greater numbers in some areas. Although the common names of these two species imply differences in habitat, environmental and biological requirements of the two species are similar. Unless otherwise indicated, crawfish will refer to these two species.

Traditionally, capture fisheries from natural waters, principally the Atchafalaya River Basin in Louisiana, have supplied the bulk of the commercial crawfish in the United States. However, production in natural waters is dependent on the amount of water in the flood basins, and thus is largely unpredictable. Because of the fluctuations in the abundance of wild crawfish, pond culture was initiated to ensure sufficient commercial supplies. In 1987, crawfish were cultured for food on approximately 54,600 hectares in Louisiana and 6000 hectares in Texas, with smaller areas under culture in other states, including Mississippi, South Carolina, and California. An estimated 30.4 million kg were harvested from culture ponds in Louisiana and Texas in 1987. Usually during 2 years out of 5, quantity of crawfish harvested from natural waters will exceed that from culture ponds.

Fig. 12.1.   Red crawfish, the predominant culture species in the United States.
(Courtesy of Edwin H. Robinson)

CULTURE

Crawfish are cultured in ponds that vary considerably in size and in dominant vegetation; however, there are two basic types of ponds, open and wooded. Open ponds are cleared areas in which the dominant vegetation may be an agricultural crop, such as rice, or aquatic and semi-aquatic weeds (Figure 12.2). Wooded ponds contain varied amounts of shrubs and trees (Figure 12.3). Ponds constructed specifically for crawfish culture are generally open earthen ponds of various sizes that retain water at levels of approximately 30 cm to 75 cm.

Culture practices range from the simplest type of culture, which involves the harvest of crawfish from natural wetlands, to intensive pond culture, where crawfish are the primary crop. Double cropping of crawfish and rice and rotation of crawfish with soybeans or rice fall within these extremes in effort expended. The present discussion will be restricted primarily to pond culture in which crawfish are the main crop.

Pond management practices vary considerably; however, a pond management scheme is used in the southern United States that follows the natural hydrological cycle. This involves stocking crawfish in the ponds in late spring, draining the ponds in early summer, reflooding in early fall, and harvesting from late fall to late spring.

Stocking.   Initial stocking of crawfish in the ponds takes place from mid to late April. Stocking rates vary from 22 to 28 kg/hectare in ponds with good vegetative cover or with native crawfish in surrounding areas, to 56 to 112 kg/hectare in ponds with little or no vegetative cover or native crawfish population. Adult crawfish should be stocked at a 1:1 ratio of males to females. Restocking is usually unnecessary in subsequent years.

Fig. 12.2. Open crawfish pond with traps.
(Courtesy of Texas Agricultural Extension Service)

Fig. 12.3. Wooded crawfish pond.
(Courtesy of Texas Agricultural Extension Service)

Draining. Crawfish ponds are slowly drained, over a period of about 2 weeks, during June. However, if rice and crawfish are double cropped (rice is planted and harvested and the field is reflooded to produce crawfish), draining may be earlier. Draining occurs near the end of breeding season when female crawfish are burrowing. Draining is necessary for vegetation to grow on the pond bottom. Rice can be planted or aquatic and semi-aquatic weeds can be allowed to grow.

Reflooding. Ponds are reflooded around September and marketable crawfish can usually be harvested by November or December. An early crop is desirable in order to market the crawfish before prices are depressed by wild-caught crawfish which usually enter the markets in April. Late flooding, after mid-October, reduces the length of the growth period prior to winter and, in addition, may result in starvation of crawfish in the burrows. Early flooding, prior to September, may result in oxygen-deficient waters due to the rapid decomposition of vegetation because of warm water temperatures.

Water quality.   The primary water quality factor limiting production is dissolved oxygen (DO). Dissolved oxygen should not drop below 3 mg/L. Methods used to prevent low DO levels include controlling the amount and type of vegetation, mechanical aeration, and water exchange. Vegetation is necessary in flooded ponds for crawfish production, but too much causes low DO problems. Huner and Barr (1984) suggest that the dry weight of standing vegetative biomass should not exceed 10,000 kg/hectare in crawfish at time of flooding. Vegetation such as aquatic plants or rice that do not die when flooded are beneficial because they do not immediately contribute to the oxygen demand generated by the microbial degradation of decaying plants.

Harvest.   Harvest of crawfish begins as soon as sufficient numbers have reached a harvestable size of approximately 75 mm in length, which is usually November/December. Early harvests consist of hold-over adults and juveniles from the previous year and fast growing young-of-the-year crawfish. Young-of-the-year crawfish can reach harvestable size within 60 to 90 days depending on water temperature and availability of food.

Crawfish are harvested by using baited traps (see Figure 12.4). Generally, 1.8-cm-mesh wire traps are used. Baits used include fish parts, beef pancreas, cottonseed cake, and commercially prepared baits. Approximately 50 traps per hectare are used. The traps are checked one or more times daily. During heavy production, April and May, traps may yield from 0.9 kg to 1.8 kg each. Trapping may continue until draining begins during June if prices remain favorable.

Yields.   Yields in static ponds may reach 1,200 kg per hectare per year, but usually average approximately 550 kg per hectare. Yields in ponds in which water is recirculated and aerated may reach 2,250 kg per hectare per year. Yields as high as 4,500 kg per hectare per year have been reported in experimental ponds. Yields from wooded or marsh ponds rarely exceed 350 kg per hectare per year.

FEEDING BEHAVIOR

Crawfish are generally considered to be detritivores; however, they are opportunistic, feeding at various trophic levels on both animals and plants as well as on detritus. Although they are indiscriminate feeders, foods of animal origin appear to be preferred. Macrovegetation constitutes the largest portion of the detritus. They also consume tissues of

Fig. 12.4. Harvesting crawfish from open pond by trapping.
(Courtesy of School of Forestry, Wildlife & Fisheries, Louisiana State University)

green macrophytes. Crawfish apparently derive much of their nutriment from the microfauna and microflora growing on the detritus and macrophytes. Crawfish are poor swimmers, but they sometimes capture food at the water surface. If natural or artificial foods are taken at the surface, the crawfish then settles to the bottom to consume the captured item.

At water temperatures above 15° C, crawfish become active feeders. In clear waters, crawfish feed primarily at night, but if the water is muddy or shaded, feeding occurs during the day. Crawfish continue to feed actively during warm weather except during periods of molting or when water temperature exceeds 30° C. Feeding stops a few days before molting but is resumed several hours after the molt, when the shell begins to harden.

Organs involved in procurement and ingestion of food are illustrated in Figure 12.5. Crawfish are attracted to food by scent, which is detected by sensory organs located on anterior appendages. Once the scent is detected, they move in the direction of the food with maxillae and maxillipeds beating to create currents that conduct the scent to the sensory organs. When contact with the food is made, via the

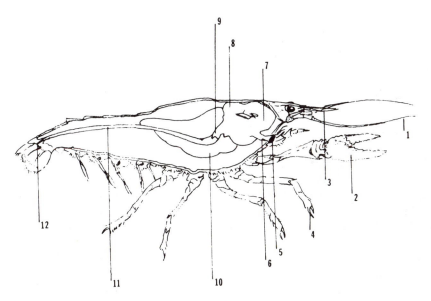

Fig. 12.5. Organs involved in procurement, ingestion, and digestion of food: (1) antenna; (2) chela; (3) antennule; (4) walking legs; (5) mouth; (6) esophogus; (7) cardiac stomach; (8) pyloric stomach; (9) midgut; (10) digestive gland; (11) intestine; (12) anus.
(Courtesy of Edwin H. Robinson)

antennae, it is seized by the chela and manipulated into the mouth by the maxillae and maxillipeds. Ingested food is shredded by the chewing mandibles. The crushed food then passes through the esophagus into the cardiac stomach. Here the gastric mill, which is composed of three chitinous teeth, grinds the food into smaller particles, of which the smallest pass through a fibrous filter into the pyloric stomach and then into the midgut. Although the pouched midgut region contains glands that secrete enzymes and absorb nutrients, most absorption occurs in the digestive gland after the food is passed into it from the midgut. Some absorption may occur in the intestine as the food passes through.

PROVIDING NATURAL FOODS FOR CRAWFISH

The major source of natural food is provided by establishing a detritus-based ecosystem in culture ponds utilizing naturally occurring vegetation or planted forages which serve as substrate for growth of food organisms. At the end of the summer growing season when the ponds are flooded, annual vegetation dies quickly and begins to decompose, providing detritus. If aquatic plants such as alligator weed (*Alternanthera philoxeroides*), water primrose (*Jussiaea* spp.), and smartweed (*Polygonum* spp.) are present, they die back to the water surface at first frost and more detritus enters the pond. Aquatic plants begin new growth in the spring. Planted rice provides detritus in fall and winter, but there is no regrowth in spring.

The nutritional value of detritus is improved as the ratio of carbon atoms to nitrogen atoms (C : N) decreases. Detritus with a C : N ratio of 17 : 1 or less is considered to be an excellent food source for crawfish (Miltner and Avault 1981). Initially, decaying organic matter has a high C : N ratio. However, as decomposition progresses, the C : N ratio decreases as a result of an increase in the microfauna population associated with the detritus. These organisms are high in nitrogen and they utilize carbon as an energy source, thus reducing the detrital C : N ratio.

Although crawfish consume detritus directly, the major source of nutrients is the detritus-based ecosystem benthic organisms which include bacteria, protozoans, small planktonic crustaceans, worms, mollusks, and insects. These organisms are good sources of the essential nutrients. Even though animal materials are important nutrient sources for crawfish, the total volume of animal material found in the stomach seldom exceeds 10% of the total food volume (Huner and Barr 1984). Green plants are second to detritus in the volume of food consumed by crawfish.

Because the nutritional value of vegetation in crawfish ponds is largely dependent on decomposition, a mixture of plants that decay at varying rates should provide food throughout the growing season. Millet, rice straw, and alligator weed is a good combination (Huner and Barr 1984). The millets decompose rapidly and provide an initial food supply at flooding, the rice straw decays slowly providing a later influx of detritus, and the alligator weed provides a late source of detritus (when killed by frost) and begins regrowth in the spring.

SUPPLEMENTAL FEEDING

A major problem in crawfish culture is depletion of food sources in early spring, which prevents the crawfish from growing quickly enough for early marketing. Supplemental feeding has been shown to increase crawfish production. Under pond culture conditions, crawfish are not fed high quality supplemental feeds for direct consumption, as are finfish. Even though crawfish readily consume pelleted fish feeds and grow rapidly, use of these feeds is usually uneconomical. Rather, supplemental feeds that enter the detrital chain and act primarily as microbial substrates are more economical than feeds designed for direct consumption by the crawfish.

Fibrous materials such as hay, rice straw, soybean stems, and leaves, corn fodder, and sugarcane wastes have been tested as supplements to existing pond food for crawfish. Production may be increased by the use of such products, but this is not always economical. Addition of hay to crawfish ponds increased production two and one-half times (Table 12.1). Hay is applied at a rate of 560 kg to 1,120 kg per hectare in Texas to supplement crawfish in ponds that have been depleted of vegetation (Davis 1984). Preliminary results have indicated that animal manures are not good for crawfish ponds because they stimulate growth of nuisance algae.

Low-protein, pelleted cattle range cubes have been evaluated in experimental and commercial ponds in Louisiana and Texas. Production is improved by their use (Table 12.2), but the economics may not be favorable (Cange, Mittner, and Avault 1982). Huner and Barr (1984) have suggested that no more than 34 kg per hectare per day of such feeds should be used at temperatures above 18° C.

The value of supplemental feeds is based on cost, on their entry into the detrital chain, and on maintaining the desirable 17:1 C:N ratio. They should be used with caution, particularly at warm temperatures, because rapid decomposition of heavy organic loads could lead to oxygen depletion.

Table 12.1. CRAWFISH PRODUCTION IN
PONDS RECEIVING HAY OR FERTILIZATION

Treatment	Average yield (kg/ha)
Rice hay	736
Bahiagrass hay	733
Fertilization	192
No fertilization or hay	202

Source: Avault et al. (1981).

Table 12.2. EFFECTS OF FEEDING CATTLE PELLETS
IN CRAWFISH PONDS WITH AND WITHOUT
PLANTED RICE

Treatment	Yield (kg/ha)	Marketable crawfish (%)
Cattle pellets only	881	5.2
Planted rice	1,274	35.7
Planted rice and pellets March–May	2,016	38.7
Planted rice and pellets September–May	2,130	48.0

Source: Cange et al. (1982).

NUTRIENT REQUIREMENTS

Although little is known about the nutrient requirements of crawfish, their qualitative nutrient requirements are presumed to be similar to those of other aquatic animals. They grow satisfactorily in closed systems on nutritionally complete commercial catfish feeds (Huner and Barr 1984), which indicates that dietary allowances for catfish feeds meet the nutritional requirements of crawfish. Data obtained with experimentally designed diets under laboratory conditions without natural foods indicated that growth rate of red crawfish was highest at protein levels of 20% to 30% and that 15% to 20% of the protein should be of animal origin (Huner and Meyers 1979). Growth and body composition data suggested that a diet containing 30% protein and 2,500 kcal/kg of gross energy is optimum for growth and protein deposition in red crawfish (Hubbard et al. 1985). It is assumed that crawfish require a dietary source of sterol as do other crustacean species. Phytosterols are as effective as cholesterol as a sterol supplement for the crawfish *Pacifacstacus leniusculus*.

Until nutritional requirements for crawfish are established, nutritionally complete diets must be formulated from known requirements of other crustaceans and finfish. Nutritionally complete diets will probably not be necessary under traditional culture practices, but will be needed for research purposes and for special culture practices outside of a pond environment.

REFERENCES

CANGE, S., M. MILTNER, and J. W. AVAULT, JR. 1982. Range pellets as supplemental crayfish feed. *Prog. Fish-Cult.* 44: 23–24.

DAVIS, J. 1984. Crawfish culture. Personal communication. Texas A&M Extension Service. College Station, TX.

HUBBARD, D. M., E. H. ROBINSON, P. B. BROWN, and W. H. DANIELS. 1985. Estimation of the protein/energy requirements of the red crawfish. *Prog. Fish Cult.* 48: 233–237.

HUNER, J. V., and J. E. BARR. 1984. Red swamp crawfish, biology and exploitation. Louisiana State University, Baton Rouge, LA.

HUNER, J. V., and S. P. MEYERS. 1979. Dietary protein requirements of the red crawfish *Procambarus clarkii* (Girard) (Decopoda Cambaridae), grown in a closed system. *Proc. World Maricul. Soc.* 10: 751–760.

MILTNER, M. R., and J. W. AVAULT, JR. 1981. An appropriate food delivery system for low-levee pond culture of *Procambarus clarkii,* the red swamp crayfish. Fifth International Symposium on Freshwater Crayfish. Davis, CA.

Appendix A:
Composition of Feed
Ingredients

The purpose of the feed ingredient composition tables (Tables A.1 through A.4) is to provide information for formulating fish feeds for research and commercial practice that can be defined nutritionally as well as prepared economically. Data in the tables come from National Research Council/National Academy of Sciences tables and other published and unpublished sources considered to be reliable by the author. Where available, each feed ingredient is assigned an International Feed Number to facilitate identification.

Values in the tables are presented on an absolute dry matter basis and must be converted to an "as fed" basis. The typical dry matter percentage of the feed is presented, but persons using the composition data are encouraged to obtain a more reliable indicator of dry matter composition of the feed they plan to use, especially moist ingredients. The values are presented on absolute dry basis because ingredients vary greatly in moisture content under different environmental conditions. In practical diet formulation, the values should be converted to an "as fed" basis.

With a few exceptions, proximate, amino acid, and mineral composition data represent chemical analyses with no correction for availability to the animal. Therefore, these values must be adjusted to allow for availability to the fish when they are used in diet formulation. This is because the nutrient requirements for fish have been determined with highly purified ingredients in which the nutrients are highly digestible and the nutrient requirement data are presented on the basis of being nearly 100% available. Available essential amino acid content is presented for several feeds based on digestibility to channel catfish. Similar amino acid availability in these feeds for other fish species is probably a safe assumption. Availability of minerals varies among sources and fish species. Availability of phosphorus in several

243

feed ingredients is presented for three fish species. Phosphorus availability in some feeds to the stomachless carp is much lower than to channel catfish or rainbow trout. Availability of minerals from technical or reagent grade chemical compounds will be higher and more consistent than from the feedstuffs.

Energy values for feedstuffs are presented on a matabolizable energy (ME) or digestible energy (DE) basis. ME values are presented for rainbow trout and DE values are presented for rainbow trout and channel catfish. Comparison of published research data indicates that cold- and warm-water species digest energy in proteins and fats relatively well and similarly, but energy in carbohydrates is more digestible to warm- than cold-water species. Thus, the DE values presented for grains, starch, and dextrin for channel catfish would probably be applicable for warm-water species, but should not be used for cold-water species.

Table A.1. PROXIMATE COMPOSITION AND ENERGY VALUES FOR PRACTICAL AND PURIFIED INGREDIENTS COMMONLY USED IN FISH FEEDS

Ingredient name	International feed number	Typical dry matter (%)	Energy, dry basis (kcal/kg)			Proximate composition, dry basis (%)			
			ME	DE		Crude protein	Crude fat	Crude fiber	Ash
			(Trout)	(Trout)	(Channel catfish)				
PRACTICAL									
Alfalfa meal	1-00-023	92.0	510	560	730	18.9	3.0	26.2	10.6
Blood meal, spray dehy.	5-00-381	93.0	3,440	3,410		93.0	1.4	1.1	7.1
Brewers grains, dehy.	5-02-141	92.0				29.4	7.2	14.4	3.9
Copra meal, solv. extr.	5-01-573	91.0				23.4	3.9	15.4	6.6
Corn distillers grain with solubles, dehy.	5-28-236	92.0				29.5	10.3	9.9	4.8
Corn distillers solubles, dehy.	5-28-237	93.0	2,280	2,440		29.7	9.2	5.0	7.8
Corn gluten meal	5-28-241	91.0	3,550	4,040		46.8	2.4	4.8	3.4
Corn, uncooked	4-02-935	89.0			1,240	10.9	4.3	2.9	1.5
Corn, extrusion cooked	4-02-935	89.0			2,840	10.9	4.3	2.9	1.5
Cotton seed meal, solv. extr.	5-01-621	91.0	2,460	2,610	2,800	45.2	1.6	13.3	7.1
Crab meal, process residue	5-01-663	92.0				34.8	2.1	11.6	44.6
Crab protein concentrate, dehy.	5-16-422	90.0				67.1	0.5		6.8
Fish solubles, condensed	5-01-969	50.0	3,350	3,680		65.3	11.2	0.9	19.2
Fish solubles, dehy.	5-01-971	93.0	4,020	4,570		69.2	8.9	1.5	13.5
Fish meal, anchovy, mech. extr.	5-01-985	92.0				71.2	4.5	1.1	16.1
Fish meal, channel catfish, mech. extr.	5-09-835	92.0				55.3			
Fish meal, herring, mech. extr.	5-02-000	92.0	4,130	4,720	4,240	78.3	9.2	0.7	11.4
Fish meal, menhaden, mech. extr.	5-02-009	92.0				66.7	10.5	1.0	20.8
Fish meal, tuna, mech. extr.	5-02-023	93.0				63.6	7.4	0.9	23.6
Fish meal, white, mech. extr.	5-02-025	91.0	2,970	3,490		68.2	5.1	0.8	25.4
Ipil-ipil (*Leucaena glauca*) leaf meal	1-16-447	92.0				29.1	6.2	12.6	9.1
Meat meal	5-00-385	94.0	3,240	3,390	3,730	54.8	9.7	2.8	28.8
Meat and bone meal	5-00-388	93.0				54.1	10.4	2.4	31.5
Molasses, sugarcane, dehy.	4-04-695	94.0				10.3	0.9	6.7	13.3
Peanut meal, mech. extr.	5-03-649	93.0				52.0	6.3	7.5	5.5

Table A.1. (continued)

Ingredient name	International feed number	Typical dry matter (%)	Energy, dry basis (kcal/kg) ME (Trout)	DE (Trout)	DE (Channel catfish)	Proximate composition, dry basis (%) Crude protein	Crude fat	Crude fiber	Ash
Peanut meal, solv. extr.	5-03-650	92.0				52.3	1.4	10.8	6.3
Poultry byproduct meal	5-03-798	93.0	2,980	3,720		62.8	14.1	2.4	16.8
Poultry, feather meal, hydrolyzed	5-03-795	93.0			3,670	91.3	3.2	1.5	3.8
Rape seed meal, solv. extr.	5-03-871	91.0	2,710	2,990		40.6	1.8	13.2	7.5
Rice bran	4-03-928	91.0				14.1	15.1	12.8	12.8
Rice middlings	1-03-941	92.0				6.8	5.6	31.5	17.1
Rice polishings	4-03-943	90.0				13.4	13.9	3.6	8.3
Shrimp meal, process residue	5-04-226	90.0				44.2	4.3	15.6	29.7
Sorghum	4-04-383	90.0				12.4	3.1	2.6	2.0
Soybean seeds, dry roasted, 204C, 12 min	5-04-597	90.0	3,840	4,190		42.2	20.0	5.6	5.1
Soybean meal, mech. extr.	5-04-600	90.0				47.7	5.3	6.6	6.7
Soybean meal, solv. extr.	5-04-604	90.0	2,570	2,980	2,870	49.9	1.4	6.5	7.0
Soybean protein concentrate	5-08-038	92.0				91.9	0.6	0.1	3.8
Sunflower meal, solv. extr.	5-04-739	93.0				49.8	3.1	12.2	8.1
Wheat bran	4-05-190	89.0			2,790	17.1	4.4	11.3	6.9
Wheat	4-05-268	88.0			2,900	14.4	1.8	2.8	1.9
Wheat flour	4-05-199	88.0				13.4	1.4	1.5	0.5
Wheat middlings	4-05-205	89.0	1,650	1,800		18.4	4.9	8.2	5.2
Yeast, brewers, dehy.	7-05-527	93.0	2,560	2,710		46.9	0.9	3.1	7.1
Yeast, torula, dehy.	7-05-534	93.0				52.7	1.7	2.4	8.3
PURIFIED									
Casein	5-01-162	91.0			4,400	92.7	0.7	0.2	2.4
Cellulose powder		96.5				0.3	Trace	96.0	
Corn starch, raw		88.0			2,700	0.3	Trace	0.1	0.1
Corn starch, cooked		88.0						0.1	0.1
20% of diet					3,400				
30% of diet					3,000				
40% of diet					2,800				
Dextrin		90.2							
30% of diet					2,920				
60% of diet					1,920				
Gelatin	5-14-503	90.0			4,700	97.4	0.1	0.0	
Glucose	4-02-891	90.0			3,380				

Table A.2. AMINO ACID COMPOSITION OF INGREDIENTS COMMONLY USED IN FISH FEEDS, BY PERCENTAGE, DRY BASIS

Ingredient name	International feed number	Typical dry matter	Crude protein	Arginine	Glycine	Histidine	Isoleucine	Leucine	Lysine	Methionine	Cystine	Phenylalanine	Tyrosine	Serine	Threonine	Tryptophan	Valine
Alfalfa meal	1-00-023	92.0	18.9	0.84	0.91	0.36	0.88	1.39	0.93	0.29	0.31	0.87	0.59	0.77	0.77	0.37	0.96
Blood meal, spray dehydrated	5-02-381	93.0	93.0	3.88	4.14	5.59	0.98	11.86	8.04	0.95	0.78	6.36	2.44	3.82	3.93	1.13	8.13
Brewers grains, dehydrated	5-02-141	92.0	29.4	1.38	1.18	0.56	1.68	2.70	0.95	0.50	0.38	1.56	1.30	1.42	1.01	0.40	1.75
Casein	5-01-162	91.0	92.7	3.85	2.77	2.86	6.32	9.71	7.88	3.10	0.34	5.31	5.41	6.03	4.32	1.19	7.40
Copra meal, solvent extracted	5-01-573	91.0	23.4	2.65	1.14	0.41	0.91	1.59	0.66	0.35	0.27	0.95	0.63		0.73	0.22	1.14
Corn distillers grains with solubles, dehydrated	5-28-236	92.0	29.5	1.05	0.55	0.70	1.52	2.43	0.77	0.54	0.32	1.64	0.76	1.42	1.01	0.19	1.63
Corn distillers solubles, dehydrated	5-28-237	93.0	29.7	1.05	1.20	0.73	1.43	2.54	0.99	0.60	0.48	1.60	0.94	1.32	1.10	0.26	1.67
Corn gluten meal	5-28-241	91.0	46.8	1.53	1.65	1.06	2.46	7.92	0.87	1.14	0.73	3.05	1.11	1.97	1.56	0.23	2.40
Corn, uncooked	4-02-935	89.0	10.9	0.48	0.42	0.29	0.39	1.37	0.28	0.19	0.25	0.54	0.43	0.57	0.40	0.90	0.50
Cotton seed meal, solvent extracted	5-01-621	91.0	45.2	4.62	2.17	1.21	1.67	2.56	1.86	0.64	0.85	2.46	1.13	1.92	1.52	0.61	2.06
				(4.19)		(1.00)	(1.20)	(1.96)	(1.32)	(0.49)		(2.05)	(0.84)		(1.16)		(1.57)
Crab meal, process residue	5-01-663	92.0	34.8	1.80	1.89	0.53	1.26	1.67	1.50	0.57	0.26	1.26	1.26	1.50	1.09	0.32	1.59
Crab protein concentrate, dehydrated	5-16-422	90.0	67.1	6.14	4.30	2.55	3.73	5.85	4.02	0.93	0.01	5.65	5.33	3.54	3.84		5.57
Fish solubles, condensed	5-01-969	50.0	65.3	3.25	7.68	2.85	2.06	3.72	3.71	1.42	0.54	2.04	0.87	2.05	1.73	0.68	2.43
Fish solubles, dehydrated	5-01-971	93.0	69.2	3.29	6.20	2.26	2.21	3.21	3.79	1.27	0.66	1.65	0.92	2.19	1.46	0.64	2.26
Fish meal, anchovy, mechanically extracted	5-01-985	92.0	71.2	4.11	4.01	1.76	3.38	5.43	5.49	2.16	0.66	3.03	2.44	2.63	3.00	0.82	3.81
Fish meal, herring, mechanically extracted	5-02-000	92.0	78.3	5.02	4.80	1.80	3.41	5.64	5.83	2.27	0.81	2.94	2.39	2.88	3.16	0.83	4.68
Fish meal, menhaden, mechanically extracted	5-02-009	92.0	66.7	4.09	4.57	1.58	3.15	4.89	5.15	1.91	0.61	2.69	2.12	2.43	2.73	0.71	3.52
				(3.71)		(1.34)	(2.73)	(4.34)	(4.43)	(1.58)		(2.34)	(1.90)		(2.38)		(3.04)
Fish meal, tuna, mechanically extracted	5-02-023	93.0	63.6	3.69	4.41	1.89	2.64	4.09	4.54	1.58	0.50	2.32	1.82	2.25	2.49	0.62	2.98
Fish meal, white, mechanically extracted	5-02-025	91.0	68.2	4.41	4.84	1.47	2.98	4.78	4.96	1.84	0.82	2.50	2.00	3.35	2.82	0.73	3.31
Gelatin	5-14-503	90.0	97.4	7.75	21.48	0.85	1.54	3.24	3.95	0.81	0.15	1.99	0.58	3.45	1.96	0.05	2.33
Ipil-ipil (Leucaena glauca) leaf meal	1-16-447	92.0	29.1														
Meat meal	5-00-385	94.0	54.8	3.84	6.71	1.02	1.86	3.40	3.45	0.75	0.70	1.94	1.02	2.30	1.75	0.37	2.68
Meat and bone meal	5-00-388	93.0	54.1	3.75	6.93	1.04	1.76	3.29	3.11	0.70	0.53	1.83	0.85	1.94	1.77	0.32	2.63
				(3.30)		(0.85)	(1.43)	(2.71)	(2.70)	(0.56)		(1.56)	(0.71)		(1.35)		(2.13)
Molasses, sugarcane, dehydrated	4-04-695	94.0	10.3														
Peanut meal, mechanically extracted	4-03-649	93.0	52.0	5.46	2.59	1.17	1.83	3.26	1.62	0.53	0.81	2.53	1.79	1.56	1.34	0.51	2.24
Peanut meal, solvent extracted	4-03-650	92.0	52.3	4.95	2.56	1.03	1.91	2.94	1.93	0.46	0.79	2.22	1.65	3.37	1.26	0.52	2.04
				(4.84)		(0.92)	(1.78)	(2.79)	(1.82)	(0.41)		(2.13)	(1.59)		(1.17)		(1.90)
Poultry by-product meal	5-03-798	93.0	62.8	4.03	5.80	1.08	2.54	4.28	3.10	1.13	0.98	1.97	1.01	2.81	2.08	0.50	3.06
Poultry feather meal, hydrolized	5-03-795	93.0	91.3	7.58	6.92	1.06	4.37	7.46	2.49	0.59	3.48	3.28	2.49	9.96	4.27	0.56	6.97
Rapeseed meal, solvent extracted	5-03-871	91.0	40.6	2.26	1.97	1.09	1.48	2.74	2.18	0.78	0.33	1.55	0.87	1.72	1.72	0.47	1.96
Rice bran	4-03-928	91.0	14.1	0.79	0.88	0.25	0.51	0.77	0.54	0.26	0.11	0.49	0.76	0.85	0.47	0.11	0.76
				(0.75)		(0.21)	(0.44)	(0.69)	(0.51)	(0.22)		(0.43)	(0.71)		(0.42)		(0.68)
Rice middlings	1-03-941	92.0	6.8	0.37	0.33	0.13	0.22	0.39	0.27	0.10	0.10	0.27	0.18	0.30	0.23	0.07	0.33
Rice polishings	4-03-943	90.0	13.4	0.57	0.78	0.19	0.39	0.78	0.58	0.14	0.14	0.42	0.46	0.54	0.38	0.11	0.88
Shrimp meal, process residue	5-04-226	90.0	44.2	2.79		1.07	1.86	2.98	2.41	0.91	0.66	1.76	1.47		1.58	0.41	2.03
Sorghum	4-04-383	90.0	12.4	0.43	0.38	0.26	0.50	1.60	0.22	0.15	0.22	0.62	0.55	0.55	0.40	0.12	0.58
Soybean seeds, dry roasted, 204C, 12 min	5-04-597	90.0	42.2	3.11	2.22	1.12	2.42	2.85	2.67	0.60	0.61	2.33	1.36	2.41	1.88	0.58	2.24
Soybean meal, mechanically extracted	5-04-600	90.0	47.7	3.41	2.64	1.26	2.92	4.02	3.10	0.72	0.63	2.45	1.72	2.23	1.92	0.68	2.53
Soybean meal, solvent extracted	5-04-604	90.0	49.9	3.38	2.03	1.19	2.27	3.65	2.99	0.58	0.83	2.36	1.48	2.36	1.85	0.71	2.25
				(3.26)		(1.04)	(1.80)	(3.03)	(2.80)	(0.49)		(1.98)	(1.23)		(1.51)		(1.77)
Soybean protein concentrate	5-08-038	92.0	91.9	8.00	3.62	2.63	5.02	6.90	6.12	0.96	1.00	4.71	3.38	5.66	3.64	0.96	4.77
Sunflower meal, solvent extracted	5-04-739	93.0	49.8	4.75	3.03	1.32	2.42	4.12	2.06	1.25	0.79	2.54	1.49	2.37	2.07	0.65	2.80
Wheat bran	4-05-190	89.0	17.1	1.09	0.97	0.44	0.57	1.03	0.65	0.22	0.36	0.62	0.48	0.77	0.51	0.28	0.78
Wheat	4-05-268	88.0	14.4	0.73	0.65	0.34	0.58	1.00	0.41	0.24	0.36	0.71	0.49	0.67	0.42	0.19	0.67
Wheat flour	4-05-199	88.0	13.4	0.49	0.51	0.28	0.53	0.99	0.28	0.21	0.35	0.69	0.39	0.68	0.37	0.14	0.57
Wheat middlings	4-05-205	89.0	18.4	1.03	0.57	0.43	0.75	1.21	0.76	0.20	0.24	0.72	0.45	0.82	0.61	0.22	0.85
				(0.98)		(0.40)	(0.66)	(1.09)	(0.73)	(0.17)		(0.67)	(0.40)		(0.54)		(0.76)
Yeast, brewers, dehydrated	7-05-527	93.0	46.9	2.35	1.87	1.17	2.37	3.45	3.33	0.79	0.53	1.96	1.60		2.27	0.55	2.52
Yeast, torula, dehydrated	7-05-534	93.0	52.7	2.83	2.85	1.42	3.06	3.78	4.01	0.83	0.65	3.06	2.14	2.96	2.83	0.56	3.17

Note: Data represent total and (available) contents. Availability is based on digestibility for channel catfish.

Table A.3. MINERAL COMPOSITION FOR MAJOR INGREDIENTS AND MINERAL SUPPLEMENTS USED IN PRACTICAL AND EXPERIMENTAL FISH FEEDS

Mineral	International feed number	Typical dry matter	Ash	Macro minerals, dry basis (%)							Micro minerals, dry basis (mg/kg)						
				Calcium	Phosphorus[1]	Potassium	Chlorine	Magnesium	Sodium	Sulfur	Cobalt	Copper	Iodine	Iron	Manganese	Selenium	Zinc
MAJOR INGREDIENTS																	
Alfalfa meal, dehydrated	1-00-023	92.0	10.6	1.52	0.25	2.60	0.52	0.32	0.11	0.24	0.33	11.0	0.16	441.0	34.0	0.37	21.0
Blood meal, spray dehydrated	5-00-381	93.0	7.1	0.52	0.26	0.10	0.27	0.24	0.42	0.37	0.01	9.0	—	2993.0	7.0	—	4.8
Brewers grain, dehydrated	5-02-141	92.0	3.9	0.33	0.55	0.90	0.17	0.16	0.23	0.32	0.08	23.0	0.07	226.0	40.0	0.76	30.0
Casein	5-01-162	91.0	2.4	0.67	0.90 (0.81)[1] (0.87)[2] (0.81)[3]	0.01	—	0.01	0.01	—	—	4.0	—	15.0	5.0	—	30.0
Copra meal, solvent extracted	5-01-573	91.0	6.6	0.19	0.66	1.63	0.03	0.36	0.04	0.37	0.14	10.0	—	750.0	72.0	—	—
Corn distillers grains with solubles, dehydrated	5-28-236	92.0	4.8	0.15	0.71	0.44	0.18	0.18	0.57	0.33	0.18	58.0	—	259.0	25.0	0.42	—
Corn distillers solubles, dehydrated	5-28-237	93.0	7.8	0.35	1.37	1.80	0.28	0.65	0.25	0.40	0.21	89.0	0.12	610.0	80.0	0.36	92.0
Corn gluten meal	5-28-241	91.0	3.4	0.16	0.50	0.03	0.07	0.06	0.10	0.39	0.09	30.0	—	423.0	8.0	1.11	190.0
Corn, uncooked	4-02-935	89.0	1.5	0.03	0.29 (0.07)[1]	0.37	0.05	0.14	0.03	0.12	0.05	4.0	—	30.0	5.0	0.08	14.0
Cottonseed meal, solvent extracted	5-01-621	91.0	7.1	0.18	1.21	1.52	0.05	0.59	0.05	0.28	0.17	22.0	—	228.0	23.0	—	68.0
Crab meal, process residue	5-01-663	92.0	44.6	15.77	1.72	0.49	1.63	1.02	0.95	0.27	0.60	35.0	0.60	4719.0	144.0	—	—
Crab protein concentrate, dehydrated	5-16-422	90.0	6.8	0.10	0.66	—	—	—	—	—	—	—	—	—	—	—	—
Fish solubles, condensed	5-01-969	50.0	19.2	0.43	1.18	3.22	5.38	0.06	4.67	0.25	0.14	92.0	2.21	445.0	27.0	3.92	87.0
Fish solubles, dry	5-01-971	93.0	13.5	1.39	1.60	0.40	—	0.32	0.40	—	—	—	—	326.0	54.0	—	83.0
Fish meal, anchovy, mechanically extracted	5-01-985	92.0	16.1	4.08	2.70 (1.08)[1]	0.78	1.08	0.27	0.95	0.84	0.19	10.0	3.41	237.0	12.0	1.47	114.0

Ingredient	Entry No.																
Fish meal, herring, mechanically extracted	5-02-000	92.0	11.4	2.40	1.82	1.17	1.08	0.16	0.66	0.50	0.06	6.0	0.57	136.0	6.0	2.07	143.0
Fish meal, menhaden, mechanically extracted	5-02-009	92.0	20.8	5.65	3.16 (1.23)[1]	0.76	0.60	0.16	0.43	0.49	0.17	12.0	1.19	524.0	37.0	2.40	162.0
Fish meal, tuna, mechanically extracted	5-02-023	93.0	23.6	8.48	4.54 (1.09)[2]	0.77	1.09	0.25	0.80	0.73	0.19	11.0	—	383.0	9.0	4.64	227.0
Fish meal, white	5-02-025	91.0	25.4	8.02	(3.36)[3] 4.17	0.91	0.55	0.20	0.85	0.53	—	6.0	—	199.0	14.0	1.77	98.0
Gelatin	5-14-503	90.0	—	0.55	(0.0–0.75)[2]	—	—	0.05	—	—	—	—	—	—	—	—	—
Ipil-ipil (Leucaena glauca) leaf meal	1-16-447	92.0	9.1	—	(2.75)[3]	—	—	—	—	—	—	—	—	—	—	—	—
Meat meal	5-00-385	94.0	28.8	9.44	4.74	0.61	1.27	0.29	1.37	0.50	0.13	10.0	—	470.0	10.0	0.47	85.0
Meat and bone meal	5-00-388	93.0	31.5	11.06	5.48	1.43	0.80	1.09	0.77	0.27	0.19	2.0	1.41	735.0	14.0	0.28	96.0
Molasses, sugarcane, dehydrated	4-04-695	94.0	13.3	1.10	0.15	3.60	—	0.47	0.20	0.46	0.46	79.0	—	250.0	57.0	—	33.0
Peanut meal, dehydrated	5-03-649	93.0	5.5	0.20	0.61	1.25	0.03	0.31	0.23	0.29	0.12	16.0	0.07	169.0	28.0	0.31	22.0
Peanut meal, mechanically extracted	5-03-650	92.0	6.3	0.29	0.68	1.23	0.03	0.17	0.08	0.33	0.12	17.0	0.07	154.0	29.0	—	22.0
Poultry by-product meal	5-03-798	93.0	16.8	3.76	1.96	0.42	0.58	0.19	0.87	0.56	0.24	15.0	3.31	473.0	12.0	0.83	129.0
Poultry, feather meal, hydrolyzed	5-03-795	93.0	3.8	0.28	0.72	0.31	0.30	0.22	0.76	1.61	0.05	7.0	0.05	81.0	14.0	0.90	74.0
Rapeseed meal, solvent extracted	5-03-871	91.0	7.5	0.67	1.04	1.36	0.11	0.60	0.10	1.25	—	—	—	—	—	1.07	—
Rice bran	4-03-928	91.0	12.8	0.08	1.70 (0.43)[2] (0.32)[3]	1.92	0.08	1.04	0.04	0.20	—	15.0	—	210.0	415.0	0.44	32.0
Rice middlings	1-03-941	92.0	17.1	0.17	0.51	0.57	—	0.11	—	0.19	—	11.0	—	249.0	10.0	—	115.0
Rice polishings	4-03-943	90.0	8.3	0.06	1.48	1.27	0.12	0.87	0.12	0.19	—	4.0	0.07	178.0	14.0	—	29.0
Shrimp meal, process residue	5-04-226	90.0	29.7	10.08	2.05	0.92	1.15	0.60	1.74	—	—	—	—	116.0	33.0	—	32.0

Table A.3. *(continued)*

Mineral	International feed number	Typical dry matter	Ash	Calcium	Phosphorus[1]	Potassium	Chlorine	Magnesium	Sodium	Sulfur	Cobalt	Copper	Iodine	Iron	Manganese	Selenium	Zinc
												Micro minerals, dry basis (mg/kg)					
Sorghum seed, dry roasted, 204C, 12 min	4-04-383	90.0	2.0	0.04	0.33	0.39	0.10	0.18	0.03	0.15	0.18	11.0	0.04	51.0	18.0	0.50	19.0
	5-04-597	90.0	5.1	0.28	0.66	1.89	—	0.23	0.03	0.24	—	18.0	—	89.0	33.0	0.12	60.0
Soybean meal, mechanically extracted	5-04-600	90.0	6.7	0.29	0.68	1.98	0.08	0.28	0.03	0.37	0.20	24.0	—	175.0	35.0	0.11	66.0
Soybean meal, solvent extracted	5-04-064	90.0	7.0	0.34	0.70	2.20	0.04	0.30	0.04	0.47	0.10	25.0	0.15	133.0	32.0	0.34	48.0
Soybean meal, without hulls, solvent extract	5-04-612	90.0	6.5	0.29	(0.35)[1] 0.70 (0.20-0.38)[1]	2.30	0.05	0.32	0.03	0.48	0.07	22.0	0.12	148.0	41.0	0.11	61.0
Soybean protein concentrate	5-08-038	92.0	3.8	0.12	0.74	0.19	0.02	0.02	0.08	0.76	0.42	15.0	0.35	149.0	6.0	0.15	37.0
Sunflower meal, solvent extract	5-04-739	93.0	8.1	0.44	0.98	1.14	0.11	0.77	0.24	—	—	4.0	—	33.0	20.0	—	—
Wheat bran	4-05-190	89.0	6.9	0.13	1.38	1.56	0.05	0.60	0.04	0.25	0.11	14.0	0.07	128.0	125.0	0.43	128.0
Wheat	4-05-268	88.0	1.9	0.05	0.43	0.49	0.06	0.13	0.02	0.15	0.14	5.0	—	35.0	33.0	0.45	43.0
Wheat flour	4-05-199	88.0	0.6	0.03	0.20	0.16	0.10	0.06	0.01	0.24	0.07	1.0	0.10	6.0	4.0	0.17	7.0
Wheat middlings	4-05-205	89.0	5.2	0.13	0.99	1.13	0.04	0.40	0.19	0.20	0.10	22.0	0.12	93.0	126.0	0.83	116.0
Yeast, brewers, dehydrated	7-05-527	93.0	7.1	0.13	(0.28)[1] 1.49	1.79	0.08	0.27	0.08	0.45	0.20	35.0	0.38	117.0	6.0	0.98	41.0
Yeast, torula, dehydrated	7-05-534	93.0	8.3	0.54	(1.39)[2] (1.36)[3] 1.71	2.04	0.02	0.18	0.04	0.59	0.03	14.0	2.69	126.0	9.0	1.08	100.0
MINERAL SUPPLEMENTS																	
Bone meal, steamed	6-00-400	96.0	28.39	28.39	13.58	0.19	0.01	0.58	0.43	0.23	—	10	—	880	30	—	440
Calcium carbonate, CaCO₃	6-01-069	100.0	39.39	39.39	0.04	0.06	—	0.05	0.06	0.09	—	—	—	300	300	—	—
Calcium phosphate monobasic (monocalcium phosphate)	6-01-082	97.0	16.40	16.40	21.60 (20.30)[1] (20.30)[2] (20.30)[3]	0.08	—	0.61	0.06	1.22	10	10	—	15,800	360	—	90

Feedstuff	No.															
Calcium phosphate, dibasic (dicalcium phosphate)	6-01-080	97.0	22.0	19.30 (12.50)[1] (8.88)[2] (13.70)[3]	0.07	—	0.59	0.05	1.14	10	10	—	14,400	300	—	100
Calcium phosphate, tribasic	6-01-084	97.0	39.20	20.1 (2.61)[2] (12.86)[3]	—	—	—	—	—	—	—	—	—	—	—	—
Cobalt acetate, tetrahydrate, $Co(CH_3 \cdot CO_2)_2 \cdot 4H_2O$	6-29-480	99.0	—	—	—	—	—	—	—	236,000	—	—	—	—	—	—
Cobalt carbonate, $CoCO_3$	6-01-566	99.0	—	—	—	0.01	—	—	0.20	460,000	—	—	500	—	—	—
Cobalt oxide, CoO	6-01-560	99.0	—	—	—	—	—	—	0.20	710,600	—	—	500	—	—	—
Copper chloride, $CuCl_2 \cdot 2H_2O$	6-01-706	99.0	—	—	—	41.6	—	—	—	—	372,700	—	—	—	—	—
Copper oxide, CuO	6-01-712	99.0	—	—	—	—	—	—	—	—	798,800	—	—	—	—	—
Copper sulfate, pentahydrate, $CuSO_4 \cdot 5H_2O$	6-01-720	100.0	—	—	—	—	—	—	12.84	—	254,500	—	—	—	—	—
Iron carbonate, $FeCO_3$	6-01-863	99.0	—	—	—	—	—	—	—	—	—	—	400,000	—	—	—
Iron fumarate, $FeC_4H_2O_4$	6-08-097	99.0	—	—	—	—	—	—	—	—	—	—	328,700	—	—	—
Iron (ferrous) sulfate, heptahydrate, $FeSO_4 \cdot 7H_2O$	6-20-734	98.0	—	—	—	—	—	—	12.35	—	—	—	218,400	—	—	—
Limestone	6-02-632	98.0	34.00	0.02	0.12	00.03	2.06	0.06	0.04	—	—	—	3,500	—	—	—
Magnesium carbonate, $MgCO_3 \cdot Mg(OH)_2$	6-02-754	98.0	0.02	—	—	0.00	30.81	—	—	—	—	—	220	—	—	—
Magnesium oxide, MgO	6-02-756	98.0	3.07	—	—	—	56.20	—	13.00	—	—	—	—	100	—	—
Magnesium sulfate, heptahydrate, $MgSO_4 \cdot 7H_2O$	6-02-758	98.0	0.02	—	—	—	9.80	—	—	—	—	—	—	—	—	—
Manganese (manganous) carbonate, $MnCO_3$	6-03-036	97.0	—	—	—	—	—	—	—	—	—	—	—	478,000	—	—
Manganese (manganous) citrate, $Mn_3(C_6H_5O_7)_2$	6-03-040	99.0	—	—	—	—	—	—	—	—	—	—	—	303,600	—	—

Table A.3. (continued)

Mineral	International feed number	Typical dry matter	Ash	Macro minerals, dry basis (%)							Micro minerals, dry basis (mg/kg)						
				Calcium	Phosphorus[1]	Potassium	Chlorine	Magnesium	Sodium	Sulfur	Cobalt	Copper	Iodine	Iron	Manganese	Selenium	Zinc
Manganese (manganous) oxide, MnO	6-03-056	99.0	—	—	—	—	—	—	—	—	—	—	—	—	774,500	—	—
Manganese sulfate, monohydrate, MnSO$_4$·H$_2$O	6-28-103	100.0	—	—	—	—	—	—	—	18.97	—	—	—	—	325,000	—	—
Oyster shell, ground	6-03-481	99.0	—	38.00	0.07	0.10	0.01	0.30	0.21	—	—	—	—	2,900	100	—	—
Phosphate, defluorinated	6-01-780	100.0	—	32.0	18.00	0.80	—	0.42	4.90	—	10	20	—	6,700	200	—	60
Potassium bicarbonate, KHCO$_3$	6-29-493	99.0	—	—	—	39.05	—	—	—	—	—	—	—	—	—	—	—
Potassium iodate, KIO$_3$	6-03-758	100.0	—	—	—	18.27	—	—	—	—	—	—	593,000	—	—	—	—
Potassium iodide, KI	6-03-759	100.0	—	—	—	21.00	—	—	—	—	—	—	681,700	—	—	—	—
Sodium chloride	6-04-152	100.0	—	—	—	—	60.66	—	39.34	—	—	—	—	—	—	—	—
Sodium phosphate, monobasic, monohydrate (monosodium phosphate), NaH$_2$PO$_4$·H$_2$O	6-04-288	97.0	—	—	22.50 (20.25)[1] (21.15)[2] (22.05)[3]	—	—	—	16.68	—	—	—	—	—	—	—	—
Sodium selenite, Na$_2$SeO$_3$	6-26-013	98.0	—	—	—	—	—	—	26.60	—	—	—	—	—	—	456,000	—
Zinc carbonate, ZnCO$_3$	6-05-549	99.0	—	—	—	—	—	—	—	—	—	—	—	—	—	—	521,500
Zinc oxide, ZnO	6-05-553	100.0	—	—	—	—	—	—	—	—	—	—	—	10	10	—	780,000
Zinc sulfate, monohydrate, ZnSO$_4$·H$_2$O	05-555	99.0	—	0.02	—	0.015	—	—	—	17.38	—	—	—	10	10	—	363,600

Note: Data represent total and (available) contents. Numbers in parentheses represent availability or absorption by various fish species.
[1] Channel catfish.
[2] Common carp.
[3] Rainbow trout.

Table A.4 FATTY ACID COMPOSITION OF THE TRIGLYCERIDE FRACTION OF INGREDIENTS USED IN FISH FEEDS

Ingredient name	International feed number	Fatty acid composition of triglyceride (%)														Sat (%)	Mono-unsat. (%)	Poly-unsat. (%)	n-6 (%)	n-3 (%)	Ratio n-3/n-6 (%)
		C12 & C14:0	C16:0	C16:1	C18:0	C18:1	C18:2	C18:3	C20:2	C20:3	C20:4	C20:5	C22:4	C22:5	C22:6						
Fish meal or oil																					
Menhaden	7-08-049	8.0	28.9	7.9	4.0	13.4	1.1	0.9	0.5	—	1.2	10.2	0.7	1.6	12.8	40.9	21.3	29.0	3.9	23.9	8.44
Herring	5-02-000	7.6	18.3	8.3	2.2	16.9	1.6	0.6	—	—	0.4	8.6	—	1.3	7.6	28.1	25.2	20.1	3.3	16.8	5.09
Salmon—sea caught		3.7	10.2	6.7	4.7	18.6	1.2	0.6	0.4	0.1	0.9	12.0	0.6	2.9	13.8	18.6	25.3	32.5	5.0	26.4	5.28
Channel catfish—																					
cultured	5-09-835	1.0	14.1	2.2	4.6	41.1	25.5	2.5	—	1.1	0.8	0.7	—	0.5	1.7	19.7	43.3	33.8	26.3	6.4	0.24
Penaeid shrimp	5-04-226	1.1	15.5	7.5	8.2	12.8	4.3	1.0	—	—	8.7	11.2	—	1.9	11.0	24.8	20.3	28.1	14.9	23.2	1.56
Animal by-product meal or fat																					
Beef	4-08-127	3.0	27.0	—	21.0	40.0	2.0	0.5	—	—	—	—	—	—	—	51.0	40.0	2.5	2.0	0.5	0.25
Pork	4-04-790	1.5	32.2	3.0	7.8	48.0	11.0	0.6	—	—	—	—	—	—	—	41.5	51.0	11.6	11.0	0.6	0.05
Grain and seed meal or oil																					
Soybean	4-07-983	—	8.5	—	3.5	17.0	54.4	7.1	—	—	—	—	—	—	—	12.0	17.0	61.5	54.4	7.1	0.13
Corn	4-07-882	—	7.0	—	2.4	45.6	45.0	0.5	—	—	—	—	—	—	—	9.4	45.6	45.5	45.0	0.5	0.01
Coconut	4-09-320	65.5	8.0	—	5.6	5.6	1.6	—	—	—	—	—	—	—	—	28.8	5.6	1.6	1.6	0.0	0.00
Cottonseed		1.0	26.0	1.0	3.0	17.5	51.5	—	—	—	—	—	—	—	—	30.0	18.5	51.5	51.5	0.0	0.00
Linseed		—	6.0	—	3.5	20.0	14.5	56.0	—	—	10.5	—	—	—	—	9.5	20.0	70.5	14.5	56.0	3.86
Canola		—	3.0	—	1.5	32.0	19.0	10.0	—	—	—	—	23.5*	—	—	4.5	55.5	39.5	29.5	10.0	0.34
Peanut		—	11.5	—	3.0	53.0	26.0	—	—	—	1.5	—	—	—	—	14.5	53.0	27.5	27.5	0.0	0.00

* 22:1, n-9

Appendix B:
Common and
Scientific Names
of Species

Common name	Scientific name
VERTEBRATES	
Japanese eel	*Anguilla japonica*
Common carp	*Cyprinus carpio*
Silver carp	*Hypophthalmichthys molitrix*
Grass carp	*Ctenopharyngodon idella*
Sea bass	*Dicentrarchus labrax*
Channel catfish	*Ictalurus punctatus*
Coho salmon	*Oncorhynchus kisutch*
Chinook salmon	*Oncorhynchus tshawytscha*
Rainbow trout	*Salmo gairdneri*
Atlantic salmon	*Salmo salar*
Brook trout	*Salvelinus frontinalus*
Blue tilapia	*Oreochromis aurea*
Mossambique tilapia	*Oreochromis mossambica*
Nile tilapia	*Oreochromis nilotica*
Other tilapias	*Tilapia zilli*
	Tilapia rendalli
INVERTEBRATES	
Brine shrimp	*Artemia salina*
American lobster	*Homarus americanus*
Freshwater shrimp	*Macrobrachium rosembergi*
Marine shrimps	*Penaeus vanami*
	Penaeus indicus
	Penaeus japonicus
	Penaeus stylirostris
	Penaeus monodon
	Penaeus serratus
	Palaemon serratus
Red crawfish	*Procambarus clarkii*

Index